THE STARS
IN THEIR COURSES

Cambridge University Press
Fetter Lane, London

Bombay, Calcutta, Madras
Toronto
Macmillan

Tokyo
Maruzen Company, Ltd.

The Origin of the Milky Way

By Tintoretto.

THE STARS
IN THEIR COURSES

BY

SIR JAMES JEANS

M.A., D.Sc., Sc.D.,
LL.D., F.R.S.

CAMBRIDGE
AT THE UNIVERSITY PRESS
1931

First Edition   *March* 1931
Reprinted   *April*   1931
Reprinted   *April*   1931

PRINTED IN GREAT BRITAIN

PREFACE

In giving a course of recent wireless talks, I assumed that my listeners had no previous scientific knowledge of any kind, and tried to introduce them to the fascination of modern astronomy and to the wonder of the universe we see through the giant telescopes of to-day.

The present book contains these talks expanded to double their original length, still in the informal conversational style and simple non-technical language of wireless talks. It is totally unambitious, aiming only at providing an easy, readable and not over-serious introduction to the most poetical of the sciences.

J. H. JEANS

DORKING

Jan. 22, 1931

CONTENTS

Chapters

Appendices

ILLUSTRATIONS

CHAPTER I

THE VAULT OF HEAVEN

We inhabitants of the earth enjoy a piece of good fortune to which we give very little thought, which, indeed, we take almost as much for granted as the air we breathe—I mean the fact that we have a transparent atmosphere. Some of the other planets, for instance Venus and Jupiter, have atmospheres which are so thick with clouds as to be totally opaque. If we had been born on Venus or Jupiter, we should have lived our lives without ever seeing through the clouds, and so should have known nothing of the beauty and poetry of the night sky, and nothing of the intellectual excitement and joy of trying to decipher the meaning of the vast panorama of lights which are scattered round us in all directions in space.

It will not form a bad approach to our subject, if we imagine that until to-night our earth had also been covered in by an opaque blanket of clouds. Suddenly this is rolled back, and we see the glory, and the tantalising puzzle, of the night-sky for the first time.

Our first impression would probably be that the stars were some sort of illumination of lamps or lanterns suspended above our heads, perhaps at only a few miles, or even yards, distance—rather like the lights in the roof of a vast tent or hall. This is what our remote ancestors thought when human intelligence began to dawn, and men first let their thoughts travel outside the earth on which they lived their daily lives.

Very soon after the cloudy curtain had been rolled back, we should notice that this array of lights was not standing still above our heads. The best way of discovering how they move is to expose a photographic plate to the sky, and let each light record its own motion. That reproduced in Plate I was exposed for $2\frac{1}{4}$ hours. Each curved line represents the path of a single star, and we see at once that the stars are going round in circles. A little observation discloses that the whole array of lights appears to turn round solid once every twenty-four hours. It is as though the lights were attached to a great hollow shell which rotated above our heads, as a telescope-dome does above the telescope. This again is what primitive man thought, and indeed civilised man also, with a few exceptions, until 300 years ago, when the discoveries of Galileo first began to reveal the true structure of the universe.

The Rotating Earth

Yet, even if we had never seen the sky before to-night, we of to-day would know that the stars do not really move in this way. Experiments which can be made on earth, without looking at the sky at all, prove conclusively that the earth turns round in space once every twenty-four hours, and so shew that it is the earth and not the sky which is rotating.

PLATE I

Norman Lockyer Observatory

The Rotating Dome of the Sky

Each curved line records the apparent path of a single star in $2\frac{1}{4}$ hours. The straight line across the plate is a meteor-trail, made by a meteor (p. 66) which happened to dash through the earth's atmosphere while the plate was being exposed.

PLATE II

Flamsteed's Atlas

Perseus and Andromeda

Andromeda is fettered to the rocks, while Perseus, carrying the Medusa's head, comes to rescue her. The star in Medusa's forehead is the variable star Algol (see p. 163). The bottom part of this plate connects with Plate III (p. 14).

The motion of the stars over our heads is as much an illusion as that of the cows, trees and churches that flash past the windows of our train.

These experiments are of two kinds. Let us discuss them in turn.

Most ships are steered by the help of an instrument known as the "magnetic compass." In this a small magnet is pivoted so that it can turn in any direction. The earth's magnetism pulls it round until it points to the north, and the navigator, knowing in this way which direction is north, steers his ship accordingly. But submarines and certain other modern ships are steered by an instrument known as the "gyroscopic compass," which works on quite a different principle. In this, a good-sized spinning-top has the ends of the axis round which it spins embedded in a frame. This again is pivoted so that it can turn in any direction. While the ship is still in port, the axis of the top is pointed north. The top is then set spinning, and is kept spinning by electrical machinery of the same kind as we use to keep an ordinary electric fan turning. No matter how much the ship turns about, the axis of the spinning-top remains always pointing to the north. The reason for this is the very simple one that there is nothing pulling on the top to change its direction of spin. Again the navigator can steer his ship by reference to this fixed direction. If his ship turns a complete circle in a fog, the compass will appear to be turning round inside the ship, and this will at once disclose that the ship has turned. When a submarine turns in a circle under the sea, its turning is shewn in the same way. And, again in the same way, a gyroscopic compass on land will shew the daily turning of the earth in space.

The turning of the earth can also be proved by a still simpler device, known as Foucault's pendulum. Try hanging a heavy weight by a long string from a high roof, and set it swinging like the pendulum of a clock. This improvised pendulum will continue to swing in the same direction in space, again for the simple reason that there is nothing to change its direction. But you will find that it will not continue to swing in the same direction in the room in which it is hanging; its direction of swing will appear to turn round in the room. The reason is that the room itself is turning round all the time in space. A careful study of the motion will shew that the earth turns round once every twenty-four hours. In many science museums and laboratories, you will find a long pendulum swinging from the roof, and by watching it for long enough, you can see the floor of the building, and with it ourselves and the whole earth, turning round under the pendulum. Just in the same way, when we watch the apparent motion of the stars over our heads, what we actually see is ourselves and the whole earth turning round under the vaulted dome of the sky. We are like children on a "merry-go-round" in a village fair. The whole fair seems to be going round them, but actually it is they who are going round inside the fair.

If we were now seeing the stars for the first time, we might reasonably think they were only a few yards or perhaps a few miles overhead. Yet we should soon find that no journey we can take over the earth would alter the directions of the stars in space, and neither would it, as a matter of fact, if the earth were hundreds of times bigger than it is, so that we could travel millions of miles from pole to pole, and we were armed with the most powerful telescope ever made. This shews how enormous the distances of the stars are in comparison with the size of the earth; our home in space, which seems so huge a globe when we travel over it, is only a tiny speck in the immensity of astronomical space.

Our nearest Neighbour—the Moon

When a journey taken over the earth's surface is found to produce a perceptible change in the direction of any object out in space, we may be sure that the object in question is nearer to us than the stars are. For instance, two observatories at different parts of the earth's surface, say Greenwich and Cape Town, cannot detect any differences in the directions of the stars, but do unmistakably see the moon in slightly different directions in space. This shews that the moon is nearer than the stars, and also makes it possible for us to estimate the distance of the moon from the earth, by a process similar to that of ordinary surveying, or of range-finding in war. We do not need to go to the top of a mountain to discover how high it is, nor need we go to the enemy's guns to find how far they are from our own. In the same way we need not go to the moon to find its distance from the earth. By this sort of surveyors' or "range-finding" method, we find that the moon is about 239,000 miles away from the earth, and, to within a few thousand miles, its distance always remains the same. Yet a very little observation shews that the moon is not standing still; while its distance from the earth remains the same, its direction continually varies. We find that it is travelling in a circle—or very nearly a circle—round the earth, going completely round once a month, or, more precisely, once every $27\frac{1}{3}$ days. It is our nearest neighbour in space, and like ourselves, is kept tied to the earth by the earth's gravitational pull, to which we shall return later (p. 69).

Apart from the sun, the moon looks the biggest object in the sky. Actually it is one of the smallest, and only looks big because it is so near to us. Its diameter is only 2160 miles, or a little more than a quarter of the diameter of the earth. Once a month, or, more exactly, once every $29\frac{1}{2}$ days, at the time we call "full-moon," its whole disc looks bright. At other times only part of it appears bright, and we invariably find that this is the part which faces towards the sun, while the part facing away from the sun appears dark. Artists could often make their pictures more convincing if they kept this in mind —only those parts of the moon which are lighted up by the sun are bright. This shews that the moon emits no light of its own. It merely reflects the light of the sun, like a huge mirror suspended in the sky.

Yet the dark part of the moon's surface is not absolutely black; generally it is just light enough for us to be able to distinguish its outline, so that we speak of seeing "the old moon in the new moon's arms." The light by which we see the old moon does not come from the sun, but from the earth. We know only too well how the surface of the sea or of snow, or even of a wet road, may reflect uncomfortably much of the sun's light on to our faces. In the same way the surface of the whole earth reflects enough of the sun's light on to the face of the moon for us to be able to see the parts of it which would otherwise be dark.

If there were any inhabitants of the moon, they would see our earth reflecting the light of the sun, again like a huge mirror suspended in the sky; they would speak of earthlight just as we speak of moonlight. "The old moon in the new moon's arms" is nothing but that part of the moon's surface on which it is night, lighted up by earthlight. In the same way, the lunar inhabitants would occasionally see part of our earth in full sunlight and the rest lighted only by moonlight; they might call this "the old earth in the new earth's arms."

The Sun

It is easy to measure the moon's distance, because it is so near the earth. It is much less easy to measure the distance of the sun, because this is much further away; the methods we use to find the moon's distance do not work very well for the sun. Somewhat, but not entirely, similar methods shew that the sun's distance is a little less than 93 million miles—probably about 92,900,000 miles. Thus the sun is about 400 times as distant as the moon, which explains why its distance is more difficult to measure.

Yet the sun and moon look about equally big in the sky. Every now and then, what is known as an "eclipse" of the sun takes place; the moon comes right in front of the sun, and is found to cover it up almost exactly. The explanation is of course that the sun is not only about 400 times as distant as the moon, but is also 400 times as big. Its diameter is about 400 times the moon's diameter, or 109 times the earth's diameter, or 864,000 miles. This of course means that the sun is 109 times as big in each direction as the earth—in length, and breadth, and height. As a consequence no fewer than 1,300,000 earths could be packed inside the sun.

The Distances of the Stars

The method I have just described will tell us the distances of the sun and moon, but it fails hopelessly if we try it on the stars. We soon find that we must take a far longer journey than from Greenwich to Cape Town before we can detect any change in the directions of the stars. Fortunately Nature herself provides such a journey and gives us free transport. The earth carries us right round the sun once a year, so that at any instant we are at the exact opposite side of the sun from our position of six months previously, and so are 186 million miles away from that position.

This journey of 186 million miles is so long that, after taking it, we do at last see the stars in slightly different directions in space, although, even so, we need extremely refined instruments to measure the change of direction. By again using the surveyors' method, this time on an incomparably larger scale, we can calculate the distance of a star from the amount by which its direction changes as we ourselves move through 186 million miles.

The distances of the nearest stars can be measured with some accuracy in this way. A very dim star far down in the Southern Hemisphere, known as Proxima Centauri[1], proves to be nearest of all. It is about 25,000,000 million miles away, so that even the nearest of the stars is about 270,000 times as distant as the sun. Although this is the nearest of all known stars, it gives out so little light that it was not discovered until quite recently. Thus it is quite possible that even nearer, but still fainter, stars may be discovered at any time. Except for the sun, moon and some of the planets (pp. 53, 55), the brightest object in the whole sky is Sirius, and this is found to be 51,000,000 million miles distant. Although it is more than twice as far away as Proxima Centauri, we receive 70,000 times as much light from it. Five other stars, besides Proxima Centauri, are known to be nearer to the earth than Sirius; as they appear fainter than Sirius, notwithstanding their nearness, they must of course be intrinsically fainter than Sirius.

[1] The significance of the names of stars is explained on p. 13, the way to identify the stars in the sky on p. 154.

The Picture-book of the Sky

Even if we were seeing the stars for the first time to-night, we should notice that they are something more than a mere random collection of points of light. There is more law and order in their arrangement than we should expect to find if bright and faint spots of light had simply been scattered at random over the face of the sky from some sort of huge pepper-pot. After we had seen the sky for a few nights, we should discover that this same ordered arrangement persisted night after night. The same groups of bright stars seen night after night would soon begin to suggest the outlines of familiar objects, which would help us to remember their arrangement on the face of the sky. It is easy to discover lines of stars, triangles, squares and letters of the alphabet, such as U, V, W, in the sky. Our ancestors, helped by vivid imaginations, saw such objects as a plough, a bear, a chair, and a serpent. In this way the stars were divided up into "constellations," or groups of associated stars.

Some of these still bear the names of common objects, but a far greater number bear the names of mythical Greek heroes and of objects occurring in Greek legends. In some cases a group of several constellations near together gives a sort of pictorial representation of a legend; the sky seems to have been utilised as a sort of permanent picture-book, and made to illustrate story after story of ancient mythology as the earth turned round under it.

For instance, six constellations near together in the sky—Cepheus, Cassiopeia, Andromeda, Perseus, Pegasus and Cetus (the sea-monster or leviathan)—illustrate the legend of Perseus and Andromeda (see Plate II, facing page 3). With the help of a description by Aratus of Soli, a minor Greek poet of the third century B.C., we may visualise the scene somewhat as follows.

Andromeda is chained by her outstretched arms to a rock in the sea. Her parents Cepheus and Cassiopeia look on from nearby, but must not help her. Cepheus has himself chained his daughter to the rocks to placate the angry gods, while Cassiopeia, whose indiscreet boasting as to her daughter's beauty had caused all the trouble, remains "seated in her stately chair" (a bright W of stars). As they impotently watch, Cetus, a sea-monster or leviathan, sent by the gods themselves, approaches Andromeda to devour her. Suddenly Perseus appears, riding the flying-horse Pegasus. He has just killed Medusa, the Gorgon, whose glance turned everything to stone; he still carries her head in his hand. Dismounting in great haste, and kicking up a cloud of dust (a crowd of very faint stars) in so doing, he presents the Medusa's head to the monster Cetus, thereby turning it to stone, and then rescues Andromeda by cutting her chains. Meanwhile the horse Pegasus is falling over backwards into a second group of constellations, all of which bear aquatic names. Besides Cetus, the sea-monster, are other fishes—Pisces (the Fishes) and Piscis Australis (the Southern Fish)—also a waterman (Aquarius), and the river Eridanus. Aratus says that the waterman has already seized Pegasus by the mane.

We shall find this group of constellations in the evening sky late in the autumn. As it sets in the west another great group appears from the east—Orion, Canis Major (the Great Dog), Canis Minor (the Little Dog), Lepus (the Hare), Monoceros (the Unicorn), and Taurus (the Bull). These shew us Orion, "the mighty hunter," wearing a dazzling belt (three bright stars in line), surrounded by his dogs and animals for the chase. He is brandishing a huge club, and is ready for Taurus, the Bull, which is rushing towards him with lowered horns (see Plate III, p. 14).

It has been suggested that yet another great group of constellations may represent some form or another of the widespread legend of the deluge; they are Argo (the Ship or Ark), Columba (the Dove), Corvus (the Raven), Lepus (the Hare), Hydra (the water-snake) and Crater (the Cup). But another interpretation is possible. For Argo was the name of the ship in which the hero Jason led his sailors, the Argonauts, on their fruitless search for the golden fleece, and the Greeks had a legend that when, after many adventures, they failed to find this, the goddess Athene changed the whole band of adventurers into stars, which now form the constellation Argo.

While most of the constellations are associated with myths and legends, one at least is associated with a historical figure. Berenice, the wife of Ptolemy III, King of Egypt, was famed for the beauty of her hair. When her husband undertook a dangerous expedition to Syria, she made a vow that if he returned in safety, she would cut off her hair and place it in the Temple of Arsinoe. In due course he returned and the queen faithfully kept her vow; she cut off her hair, and handed it to the priest in charge of the temple. As this was before bobbed hair had become fashionable, the king was exceedingly angry. To smooth matters over, the wily priest explained that the hair had already been deposited in the heavens, where its beauty would be seen by all men for ever; he pointed to a group of stars which certainly look somewhat like hair, and have been known as Coma Berenices (Berenice's Hair) ever since. So if you want to see how beautiful the tresses of the Egyptian queen were, you need only look up at the sky any spring evening, at no great distance from the Plough or Great Bear, and there they are still shining in all their glory.

Star-names

When we want to find a house in a town, we first inquire what street it is in. In the same way, if we want to find a star in the sky, we first ask to what constellation it belongs. While some houses in a town are merely identified by a number and the name of a street—27 High Street, for instance—the more conspicuous houses may have individual names of their own. It is the same with the stars; the brightest and best known have their individual names—Sirius, Arcturus, Capella, Vega, and so on—while others have only a number and a constellation, such as 27 Canis Majoris (No. 27 of the Big Dog). But before afflicting the stars with the indignity of mere numbers, astronomers first use up the letters of the Greek alphabet—Alpha (α), Beta (β), Gamma (γ), Delta (δ), Epsilon (ϵ), and so on—so that the principal star in a constellation, which is generally also the brightest, is described as Alpha of the constellation; the second, generally the second brightest, as Beta; the third as Gamma, and so on. For instance the brightest star in the whole sky may be identified either by its individual name, which is Sirius (meaning "sparkling"), or by what we may call its "constellation address," Alpha Canis Majoris, which indicates that it is the brightest star in the constellation of the Big Dog. For this reason Sirius is known as the dog-star.

The faintest stars of all do not even have constellation addresses. To identify them, we must mention their exact position in the sky, or possibly their number in some star-catalogue. For instance Wolf 359 means star number 359 in the catalogue of the astronomer Wolf.

A list of the twenty stars which look brightest in the whole sky, together with their constellation addresses, will be found in Appendix II (p. 181).

The Pole Star

Look fairly well up in the northern sky on any clear night, and you will see four fairly bright stars forming an oblong figure, slightly bashed-in at one corner. From this corner runs out a slightly curved line of three stars. The last of the three is the "Pole Star," round which the whole vault of the sky appears

PLATE III

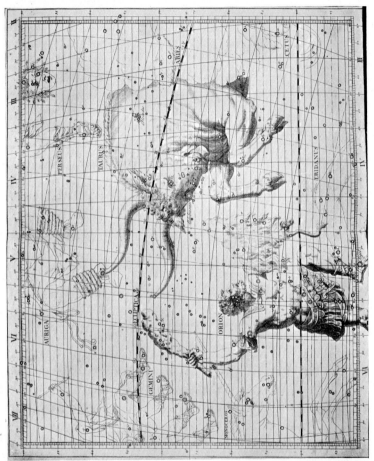

Flamsteed's Atlas

Orion and the adjoining Constellations

Orion is preparing to meet the onslaught of Taurus, the Bull (see pp. 12 and 170). The thick line just above Orion's Belt is the Equator; the thick line which goes between the horns of the Bull is the Ecliptic, the path of the sun through the sky. The top left hand is continued on Plate II.

PLATE IV

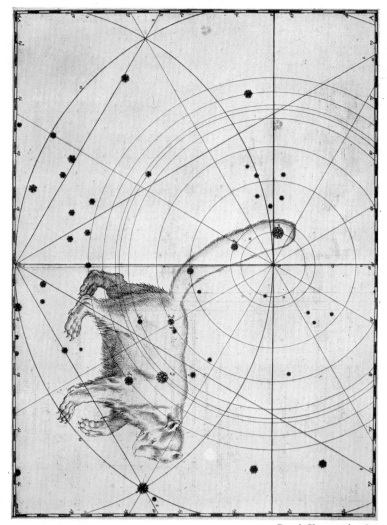

Bayer's Uranometria, 1603

The Little Bear and the Pole Star

The Pole Star is the bright star at the end of the Bear's tail. The true pole does not coincide exactly with this star, but is just underneath where the lines and circle meet. It is gradually moving along this circle—5000 years ago it was in front of the Bear's nose (see p. 18).

to turn. These seven stars, together with innumerable fainter stars, make up the constellation of Ursa Minor (the Little Bear); the oblong is its body, and the three other stars form its tail, the Pole Star being the tip of its tail (see Plate IV). It is as though the unfortunate Little Bear was tied by the tip of its tail, and swung across the sky, from east to west; indeed, the whole sky moves as though it were pivoted on the tail of the Little Bear, and swung round so as to complete the circle once every twenty-four hours.

Ranged around the Pole Star and the Little Bear are all the constellations which are most familiar to us—Ursa Major (the Great Bear), Cassiopeia, Perseus, Camelopardalis (the Cameleopard), and Draco (the Dragon) (see Star-Map I at end of book). These are familiar to us because they never set; they are visible at all times of the night and at all seasons of the year. Further away from the Pole Star than these, are other constellations which are not visible all the time, such as Orion, Canis Major (the Big Dog), Hydra, Leo (the Lion), Hercules, Serpens (the Serpent), Aquila (the Eagle), Cygnus (the Swan), Capricornus (the Goat) and Pegasus. These rise in the east at their appointed times, cross the sky until they set in the west, and then become invisible until they rise again the next night (see Star-Maps I and II at end of book). And, still further away from the Pole Star, are yet other constellations which we in England never see at all, until we travel to countries far south; such are the Southern Cross, the Centaur, the Ship, the Clock and the Table.

The Wanderings of the Pole

Prolonged observation shews that the panorama of the constellations not only comes round unchanged night after night, but remains the same from year to year, and even from generation to generation. Indeed ancient star-maps shew that the arrangement of the constellations looks practically the same to us as it did to the Egyptians, the Chinese and the Chaldeans, when first they began to study the face of the sky 5000 years ago.

Yet in one important respect, the sky looked very different to them. We see it turning night after night round the tip of the tail of the Little Bear; the astronomers of 5000 years ago saw the same sky and the same constellations turn round the star Thuban, or Alpha Draconis—a bright star in the constellation of the Dragon. This lies half-way along the tail of the Dragon, and is also in the position of a bun held in front of the nose of the Little Bear (see Plate IV, p. 15).

It may at first seem very mysterious that the pivot of the sky should wander about like this, and yet there is a very simple explanation. The vault of heaven seems to be pivoted on the Pole Star because the earth rotates about an axis which points to the Pole Star. Now the earth is, to all effects and purposes, a huge spinning-top suspended in space. When we were discussing the "gyroscopic compass," we saw how the axis of a spinning-top would always point in the same direction in space, unless something intervened to alter that direction. If then, the earth's axis is all the time changing its direction in space, something must be intervening all the time to produce the change. And we know now what it is.

We shall see below (p. 72) how the earth is firmly held by the sun's gravitational grip, and, as a consequence, moves round the sun once a year. If the earth were strictly globular in shape, the sun's gravitational grip would have no effect on it beyond preventing it running off into space. But, as it happens, the earth is not quite globular—it is rather orange-shaped, bulging a bit at its equator. And the sun's gravitational pull on this bulge slowly but continually changes the direction in which the earth's axis points in space. The result is that the pole of the heavens— the region of the sky to which the earth's axis points —moves round in the sky in a circle which takes 25,800 years to complete. This phenomenon is known as "precession."

This is not quite all, for the moon also exerts a gravitational pull on the earth, and this adds a small rapid wobble, known as "nutation," to the slower and more stately motion produced by the gravitational pull of the sun.

Because of these motions the earth's axis pointed in a different direction in the past from that of to-day, and our ancestors of 5000 years ago saw the heavens turning round a point in the constellation of the Dragon. For the same reasons, our posterity of 5000 years hence will see the heavens turning round a point in the constellation of Cepheus. The wanderings of the pole—i.e. of the earth's axis of rotation—are shewn in Fig. 1.

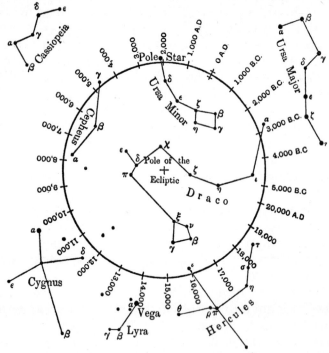

Fig. 1. *The Wanderings of the Pole.* This shews the positions occupied by the pole at various dates. Even 3000 years ago the pole was 17 degrees further south than now, so that Europeans could see parts of the southern sky which are now invisible. This explains why many of the southern constellations have Greek and Latin names.

Yet the general arrangement of the stars in the sky was the same 5000 years ago as now, and will be the same 5000 years hence; only our own little earth has changed, and not the distant stars. But, although thousands of years produce no perceptible change in the general arrangement of stars in the sky, a few of the brightest spots of light in the whole sky change their positions quite rapidly. These are called planets from the original Greek name, which meant "wanderers." They are the gipsies of the sky. They have no constellation address, just as caravan-dwellers have no postal address, and can have none because they move on from day to day.

Planets

Five planets were known to the ancients—Mercury, Venus, Mars, Jupiter and Saturn—and of course they did not know that the earth itself was a sixth. Three much fainter planets have been discovered in recent times—Uranus in 1781, Neptune in 1846 and Pluto in 1930.

We can generally detect the motion of an aeroplane overhead in a very few seconds, and the nearer to us it is, the quicker we are to notice its motion. The speeds with which astronomical bodies move are far greater than that of any aeroplane, generally many thousands of times greater. For the moment we may, without very serious error, regard them as all being the same. This being so, the speed with which any body appears to move across the face of the sky will give a rough indication of its nearness—the quicker it seems to move the nearer it must be. One exception must be mentioned—the moon. We do not see its true motion in space because it is journeying through space with us; it is, so to say, travelling in the same railway carriage as ourselves.

The photograph reproduced in Plate V illustrates the general principle in terms of two very extreme instances. The slant streak across the lower half of the plate is the path of a meteor (see p. 65) which moved across the sky so rapidly that it crossed the whole plate in a fifth of a second, a minute fraction of the whole time of exposure. The big object near the centre of the plate is the Great Nebula in Andromeda (see p. 122). This moves across the sky so slowly that it will hardly have changed its position in a million years. Both the meteor and the nebula were moving with thousands of times the speed of an aeroplane. But the meteor was near, about 50 miles up in the earth's atmosphere, and so appeared to move rapidly, while the nebula, at a distance of about 5,300,000,000,000,000,000 miles, appeared to move very slowly.

The other objects shewn in the photograph are fairly bright stars at distances intermediate between those of the two objects just mentioned. These, too, are moving through space thousands of times faster than an aeroplane. They are nothing like so far away as the nebula, and yet their distance is such that it takes thousands of years of motion at this terrific speed, before we can see that a star has changed its place in the sky.

Astronomers have a very simple device for detecting planets and other bodies whose motion across the sky is so rapid as to be easily noticeable. If a group of people are being photographed with a fairly

PLATE V

National Observatory, Prague, Czecho-Slovakia

A Fire-ball and a Nebula

The fire-ball (a big meteor) crossed the plate in a fifth of a second; the nebula
hardly moves appreciably in a million years (p. 20).

PLATE VI

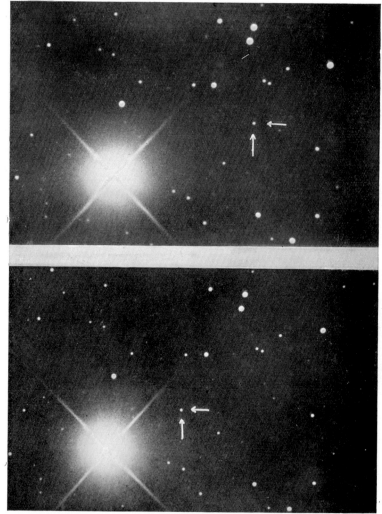

Lowell Observatory

The Discovery of Pluto

Two photographs of the region near δ Geminorum taken on March 2 and March 5, 1930. The object marked by the arrows was found to have moved considerably in the three days interval, and this established that it was of planetary character.

long exposure, and one of them moves while the plate is being exposed, the picture will be spoilt; the culprit will not figure in it as a human being, but as a fuzzy, blurred image. Taking advantage of this, the astronomer photographs a bit of the sky with a long exposure, and any object which is moving rapidly across the sky appears as a blurred image, instead of as a sharp point of light. It is by variations and modifications of this simple device that many of the objects which are near to us in space have been detected, including the remote planet Pluto, which, after many years' search, was only discovered in March, 1930 (p. 74). Plate VI shews two plates of the region of the sky in which the new planet was supposed to lie, taken at the Lowell Observatory, Arizona, at an interval of three days apart. The object indicated by arrows is seen to have moved appreciably in the interval, and this motion establishes its planetary character.

An Isolated Colony

We might perhaps have expected to find a gradual transition from these rapidly moving objects to the slowly moving stars, which shew so little sign of motion that we describe them as "fixed stars." But we do not. There are two distinct classes of objects, with nothing in between. There is a very simple reason for this. Our earth belongs to a small colony which is almost completely isolated in space, so that all the planets and other objects which belong to it are enormously nearer than even the nearest of the fixed stars. They seem to move rapidly purely and simply because they are near to us, and not because they are moving at a great number of miles an hour—actually most of them cover fewer miles in an hour than the fixed stars. Even the nearest of the stars is about 270,000 times further away than the sun, and so is about 7000 times as distant as Pluto, the most distant planet yet discovered. The light from Pluto takes between four and five hours to reach us, but the light from even the nearest of the stars takes between four and five years. This shews vividly how completely isolated this little colony is in space. Its isolation far exceeds that of human settlements in the wildest countries of the earth. We call ordinary human settlements isolated if they are several miles apart, but if we think of this colony of bodies in space as a small village in England, then the next settlement—the nearest of the fixed stars—must, on the same scale, be placed somewhere in Africa or Siberia.

The principal member of this isolated colony is of course the sun. We may properly think of it as a huge planet, although one which is far and away larger and brighter than any of the others. Like the other planets, it continually moves across the background of constellations formed by the enormously more distant fixed stars. We do not ordinarily notice this motion, because the sun's light blots out the light of all the other stars. Yet the astronomer who can see the stars through his telescope even in full daylight can follow it quite easily, and, indeed, it can be verified indirectly without a telescope. The sun is in the south at midday, and so is in the exactly opposite direction of space (north, but below the horizon) at midnight. If we look south night after night at exact midnight, we shall find that a different part of the sky lies due south each night, and this of course means that the sun, which is exactly opposite, is also in a different part of the sky each night.

Until the Middle Ages, it had been generally, although not universally, supposed that the earth formed the centre of this colony of objects, and indeed of the whole universe. It was believed that the sun, moon and planets were attached to transparent spheres which revolved round the earth at different distances, while the fixed stars were attached to a larger sphere which revolved round the central earth at a greater distance, and so formed a background to them all. Then in 1555 Copernicus published his great work, *De revolutionibus orbium coelestium,* in which he shewed that the observed motions of the sun and planets were far more simply explained by supposing that the earth was merely a planet like the others, and that all the planets, including the earth, revolved round a fixed central sun. To most minds this remained little more than a conjecture until its truth was proved by the telescopic observations of Galileo and his successors. But it is now established beyond doubt that the sun, and not the earth, forms the centre of our little colony in space, and that the earth, like all the other smaller members of the colony, goes round the central sun.

CHAPTER II

A PRELIMINARY JOURNEY THROUGH SPACE AND TIME

We cannot ourselves go and find out what the sun, moon or stars consist of, but our huge telescopes will, in a sense, bring them near to us, which comes to much the same thing. Thus the whole of space lies open for our exploration, at any rate until we are confronted by opaque substances which no telescope can penetrate. Even then the calculations of the mathematician are ready to carry on the story; for instance, quite a lot of work has been done in recent years on the constitution of the interiors of the stars. Telescopic observation and mathematical theory between them furnish us with a sort of magic rocket which will take us almost anywhere in space we desire to go.

Out in Space

Let us enter this magic rocket and persuade someone to shoot us towards the sun. We need only start with speed enough to carry us a short distance away from the earth—about 7 miles a second will do—and the sun's huge gravitational pull will do the rest. It will drag us down into the sun whether we like it or not. If we start at 7 miles a second, the whole journey will take about ten weeks.

Even in the first few seconds of our flight, we notice strange changes; the whole colour-scheme of the universe alters with startling suddenness. The sky rapidly darkens in hue, until finally it assumes a blackness like that of midnight, from which the stars shine out. They no longer twinkle in the friendly way we are accustomed to on earth; their rays have become piercing needles of steady light. Meantime the sun has changed to a hard steely whiteness, and the shadows it casts are harsh and fierce. Nature seems to have lost a large part of her beauty, and all of her softness, in a surprisingly short space of time. The explanation is that a very few seconds take us entirely clear of the earth's atmosphere, and not until we have left it behind us do we realise how much its softening effect has added to the pleasure of our lives.

Let us pause for a moment to consider the scientific reasons for this. Imagine that we stand on any ordinary seaside pier, and watch the waves rolling in and striking against the iron columns of the pier. Large waves pay very little attention to the columns —they divide right and left and re-unite after passing each column, much as a regiment of soldiers would if a tree stood in their road; it is almost as though the columns had not been there. But the short waves and ripples find the columns of the pier a much more formidable obstacle. When the short waves impinge on the columns, they are reflected back and spread as new ripples in all directions. To use the technical term, they are "scattered." The obstacle provided by the iron columns hardly affects the long waves at all, but scatters the short ripples.

We have been watching a sort of working model of the way in which sunlight struggles through the earth's atmosphere. Between us on earth and outer space the atmosphere interposes innumerable obstacles in the form of molecules of air, tiny droplets of water, and small particles of dust. These are represented by the columns of the pier.

The waves of the sea represent the sunlight. We know that sunlight is a blend of lights of many colours —as we can prove for ourselves by passing it through a prism, or even through a jug of water, or as Nature demonstrates to us when she passes it through the raindrops of a summer shower and produces a rainbow. We also know that light consists of waves, and that the different colours of light are produced by waves of different lengths, red light by long waves and blue light by short waves. The mixture of waves which constitutes sunlight has to struggle through the obstacles it meets in the atmosphere, just as the mixture of waves at the seaside has to struggle past the columns of the pier. And these obstacles treat the light-waves much as the columns of the pier treat the sea-waves. The long waves which constitute red light are hardly affected, but the short waves which constitute blue light are scattered in all directions.

Thus, the different constituents of sunlight are treated in different ways as they struggle through the earth's atmosphere. A wave of blue light may be scattered by a dust particle, and turned out of its course. After a time a second dust particle again turns it out of its course, and so on, until finally it enters our eyes by a path as zigzag as that of a flash of lightning. Consequently the blue waves of the sunlight enter our eyes from all directions. And that is why the sky looks blue. But the red waves come straight at us, undeterred by atmospheric obstacles, and enter our eyes directly. When we look towards the sun, we see it mainly by these red rays. They are not the whole light of the sun; they are what remains after a good deal of blue has already been filtered out by atmospheric obstacles. This filtering of course makes the sunlight redder than it was before it entered our atmosphere. The more obstacles the sunlight meets, the more the blue is extracted from it, and so the redder the sun looks. This explains why the sun looks unusually red when we see it through a fog or a cloud of steam. It also explains why the sun looks specially red at sunrise or sunset—the sun's light, coming to us in a very slantwise direction, has to thread its way past a great number of obstacles to reach us. It also explains the magnificent sunsets which are often seen through the smoky and dusty air of a city—or even better after a volcanic eruption, when the whole atmosphere of the world may be full of minute particles of volcanic dust.

In such ways as this, the earth's atmosphere breaks up the sunlight. The true sunlight, as it is when it leaves the sun, or travels through space before meeting the earth at all, is a blend of all the colours into which the earth's atmosphere breaks it up. To reconstruct this colour, we must blend the blue of the sky with the yellow or red of the direct sunlight. This makes the steely-white light we see as soon as our rocket takes us beyond the earth's atmosphere.

This action of the atmosphere in breaking up sunlight is responsible for much of the beauty of the earth—the blue sky of full day, the vivid orange and red of the rising and setting sun, the fairyland hues of the clouds at sunrise and sunset, the mysterious tones of twilight, the pink afterglow on the mountains, the purple of the distant hills, the apple-green in the western evening sky and the indigo in the east, and indeed all the effects which the artist describes as atmospheric. As we pass beyond the earth's atmosphere, we leave all these behind us, and enter a hard world which is divided sharply into light and dark, and knows nothing of half-tones. For the first time in our lives, we see the sun for what it really is—a vivid bluish globe of light. We see it set in a sky as black as that of midnight, because the earth's atmosphere no longer takes its rays and scatters them in all directions. It is to this weird and terrifying object that our rocket is taking us.

A close view of the Moon

If we are wise we shall have started sometime near the time of new moon, because then our path will take us near to the moon, and we can study it from close quarters. Down behind us, the surface of the earth looks murky and blurred; we see it through a thick layer of air, dust, fog and clouds, with rain and snow here and there. By comparison, the moon looks strangely clear and sharp cut. The reason is that it has no atmosphere, and as a consequence no rain, fog, clouds or dust to interfere with our vision.

Even from afar, we can see that there is no water on the moon. If there were seas, lakes, or even rivers,

PLATE VII

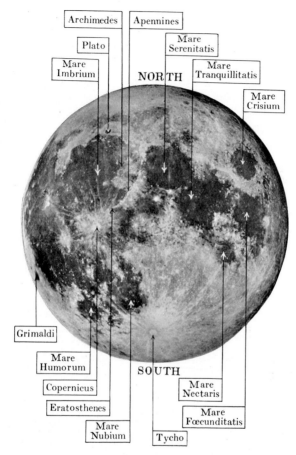

Archimedes Apennines Plato Mare Serenitatis Mare Imbrium NORTH Mare Tranquillitatis Mare Crisium Grimaldi Mare Humorum Copernicus SOUTH Mare Nectaris Eratosthenes Mare Fœcunditatis Mare Nubium Tycho

The Moon

This plate shews the full moon as seen by the unaided eye or through field-glasses (an astronomical telescope inverts all objects).

If the plate is viewed from 9 yards distance, the moon will look the same size as the real moon in the sky, and it will be possible to recognise the "Man in the Moon," the "woman reading a book," the "old man with his bundle of sticks," and so on.

PLATE VIII

Paris Observatory

The Moon (third quarter)

Here the moon is 21 days old, so that it is $8\frac{1}{2}$ days until new moon, and the moon is decreasing in size. The sun is of course to the left of the moon, so that the moon is preceding the sun (by about 7 hours) through the sky.

PLATE IX

Paris Observatory

The Moon (first quarter)

Here the moon is rather less than 6 days old, so that it is increasing in size. The sun is now to the right, so that the moon is following the sun (by about 5 hours) through the sky.

PLATE X

Lunar Detail. The region round Copernicus (see Plate VII)

The crater of Copernicus, at the centre of the plate, is 50 miles in diameter.
Eratosthenes, 35 miles in diameter, lies slightly higher to the right.

PLATE XI

Mount Wilson Observatory

Lunar Detail. The Mare Imbrium (see Plate VII) and the
surrounding mountains

From Eratosthenes (cf. Plate X) near the bottom left-hand corner, the lunar Apennines
curve round the right-hand edge of the plate. The big crater near the top is Plato,
below it is Archimedes.

PLATE XII

Craters and volcanic formations

This is not lunar scenery, but the terrestrial volcano Vesuvius and the country to the west, photographed from a model made by James Nasmyth, the engineer. It may be compared with the lunar scenery shewn on Plates X and XI.

we should be sure to see them glittering in the vivid sunlight; there is no trace of anything which in the least resembles an expanse of water. And as we get nearer, we see that there are neither cities, nor fields, nor forests. We are looking on a dead world.

Ninety-five years ago, a New York newspaper perpetrated what was afterwards known as the "great moon hoax." It published a series of wholly fraudulent articles, which claimed to describe the moon as seen through a giant new telescope in South Africa. They described trees of amazing growth, weird animals, and flying men, all of types utterly different from anything known on earth. These articles so far increased the circulation of a little-known newspaper, that it claimed to have the largest circulation of any paper in the world—a sensational proof of the interest men feel in the problem of life in other worlds.

From our rocket, we look out on a picture very different from that drawn by the American newspaper. We see that the surface of the moon consists largely of vast flat deserts, shewing no signs of cultivation or life of any kind. Scattered over the greater part of it are circular elevations which look like the rims of craters of extinct volcanoes, which is what they probably are (see Plate XII). Many of these are large enough to contain a whole English county inside. Four are larger than Devonshire, while the largest of all, Maurolycus, would just include the whole of Wales. Here and there we see immense jagged peaks and ranges of mountains, as sharp cut as when they first came into existence. The mountains on our earth have been weathered by millions of years of snow, rain and wind, but we see no trace of weathering here. If ever rocket-travelling through space becomes common, it looks as though these mountains would form a perfect paradise for climbers. The sun casts shadows of their jagged outlines on the flat deserts below, and even in a small telescope one can see wonderful needles, pinnacles and arêtes. One range of lunar mountains, the Apennines (shewn slanting across the lower half of Plate XI) is about 450 miles in length and comprises over 3000 peaks. The highest of these, Mount Huyghens, is about 19,000 feet high, while two others, Mount Bradley and Mount Hadley, are over 15,000 feet in height. To the north of the range is a flat plain (Mare Imbrium) to which the mountains fall almost precipitately, like a line of cliffs at the seashore.

The lunar mountains have other attractions, besides scenery, to offer to climbers. On the moon the force of gravity is only a sixth as great as on earth, so that a man could jump six times as high as on earth, could climb six times as high without getting tired, and could fall six times as far without getting hurt. Yet, because the moon has no atmosphere, climbers must remember to take supplies of oxygen with them.

The feebleness of gravity on the moon explains why the moon has no atmosphere. Our rocket was only able to jump right clear of the earth because we started with the high speed of 7 miles a second—if we had started out with any smaller speed we should have fallen back again onto the earth, just as an ordinary shot from a gun does, or a cricket ball driven upwards off a bat. The earth's atmosphere consists of millions of millions of molecules darting about with quite high speeds—hundreds of yards, and even miles, a second. But they never attain the speed of 7 miles a second which would take them clear of the earth, so that they continually fall back like the cricket ball, and the earth retains its atmosphere.

Against this, a projectile needs a speed of only $1\frac{1}{2}$ miles a second to get clear of the moon and right away into space; if once it gets off into space with a speed as high as this, the moon's gravitational pull is too feeble to draw it back. Now the moon always turns the same face to the earth, and only goes round it once a month. It follows that the moon turns round in space once a month, so that after any region of its surface has once got into the sunlight, it stays there to be baked for a whole fortnight. As a result, it gets very hot indeed, its temperature rising to somewhere in the neighbourhood of 200° Fahrenheit, which is but little below the temperature of boiling water. If ever the moon had an atmosphere, its molecules must have acquired high speeds of motion in this extreme heat. Calculation shews that they would frequently exceed the critical speed of $1\frac{1}{2}$ miles a second, and fly off into space never to return. And this is the whole story of how the moon lost its atmosphere.

Although the moon may, at first sight, look a paradise for climbers, mature consideration suggests that it may after all be but little suited either for a holiday resort or a permanent abode. A holiday party would not only have to take their own supplies of oxygen with them, but would also have to go prepared for a temperature of about 200° Fahrenheit on the sunny side—indeed, directly under the sun it may be as high as 244° Fahrenheit above zero, or 32 degrees above the temperature of boiling water. If this proves too hot, the only alternative is the shady side, where things are worse, the temperature being about 244 degrees *below* zero—or of course coming back home.

What is the Moon made of?

Further than this, the surface of the moon is hardly such as to make a comfortable camping ground. M. Lyot of Meudon has recently examined how ordinary moonlight, which is of course sunlight reflected from the surface of the moon, compares in quality with sunlight reflected from various kinds of soils, clays, chalks and rocks. He found that it could be matched almost perfectly by the light reflected from volcanic ash, but could not be matched at all by the light reflected from any of the many other kinds of substance he tried. This makes it highly probable that the moon's surface consists of some sort of volcanic ash, and certainly this is in keeping with the general appearance of the lunar scenery, which looks exactly like a vast exhibition of extinct volcanoes. Indeed these volcanoes are surprisingly similar to volcanoes on earth, as we have already seen in Plate XII (p. 29), which is a photograph of a model of Vesuvius and the volcanic country to the west of it.

Volcanic ash has the remarkable property of being an almost perfect non-conductor of heat, like the asbestos which is used for lagging hot-water pipes. If the moon's outer surface really consists of this substance, the heat which the sun pours down on its sunny side will not sink far in, so that the moon's interior will not experience the same violent changes of temperature as its surface. Calculation shews that the surface which has been baked in the sun for a fortnight may reach the temperature of boiling water, while the rock even half an inch below the surface is still below the freezing-point. Just as half an inch of asbestos prevents the heat escaping from our hot-water pipes, so half an inch of volcanic ash would prevent the sun's heat penetrating to the moon's interior. This is not pure imagination; it probably describes the actual state of things on the moon fairly well. Two Mount Wilson astronomers, Pettit and Nicholson, have recently recorded the changes of temperature of the moon's surface during the progress of an eclipse. They found that as the earth's shadow crossed the face of the moon, and so cut off the supply of heat from the sun, the temperature fell, quite suddenly, from 194° Fahrenheit to −152°, or 184 degrees of frost—a drop of 346 degrees in a few minutes! We are accustomed to fairly marked changes of temperature on earth at an eclipse of the sun, and generally begin to feel very chilly as soon as the moon's shadow suddenly cuts off our supply of sunshine, but we never experience anything at all comparable to this. The reason is that the store of heat in our soil and atmosphere prevents the temperature from changing very abruptly. The dramatic suddenness with which the moon's surface changes from hot to cold shews that it possesses no store of heat at all comparable with that in the soil of the earth. This in turn means that the sun's heat can only penetrate a very thin surface-layer of the moon, and the rapidity with which the lunar temperature changes is entirely consistent with the supposition that the lunar surface consists of volcanic ash.

Venus and Mercury

Clearly then the moon is no place for a prolonged stay, and we had better let our rocket carry us on to the sun, as we originally intended. After the moon, our nearest neighbour in space is the planet Venus. If we should happen to pass near it on our journey we should see nothing specially interesting. It is merely a globe about as large as the earth, completely enveloped in clouds.

But the next planet, Mercury, ought to provide an arresting spectacle. It is considerably smaller than the earth; sixteen Mercuries rolled into one would barely make one earth. Indeed it is not much larger than the moon, and like the moon, it has no atmosphere, again because its gravitational pull is not strong enough to retain one, so that its scenery should stand out in vivid relief. It is like the moon in still another respect. The moon is so tightly held in the earth's gravitational grip that it cannot rotate in this grip, and so always presents the same face to the earth. Mercury is in a similar situation; it is so tightly held in the gravitational grip of the sun that it always presents the same face to the sun. We have seen how the moon's face gets thoroughly hot after being baked in sunshine for a fortnight at a time. That hemisphere of Mercury which faces the sun is in a much worse plight; it is baked for ever and ever in the rays of the very much nearer sun, and so must be terrifically hot. If there are any rivers on it, they must be rivers of molten lead or some similar substance, since the heat is such that all ordinary liquids would boil away. There is yet one other respect in which Mercury resembles the moon. The light reflected from its surface can again only be matched by light reflected from volcanic ash, so that it seems likely that the surface of Mercury, like that of the moon, consists of this substance. Very possibly its landscape too consists of extinct volcanoes, although our rocket does not take us near enough to see whether this is so or not.

Outside the Sun

We are now well advanced on our journey to the sun. Even as we pass Mercury, it looks seven times as big as it did when we started from earth, and as we approach still nearer, and it fills the greater part of the sky in front of us, we begin to get a fine view of its surface. Clearly the sun is no dead world, like the moon and Mercury. On the contrary, we see nothing at rest; everything is in violent motion; the whole surface is agitated, boiling and erupting in various ways. We can understand why this must be. The interior of the sun is a huge power station which works continuously. The energy which is generated and set loose in the interior makes it terrifically hot, so that a vast stream of heat is driven outwards to the surface, whence it pours away into space in the form of radiation. Every square inch of surface receives fifty horse-power of energy to get rid of somehow or other. And it cannot do this by simply lying still at rest. Everywhere we see it boiling up—the topmost layers, so to speak, turning over and presenting their hottest sides to outer space, so that the imprisoned radiation can pour away all the more rapidly (see Plate XIII).

Even this is not enough, for here and there huge fountains of flame, called "prominences," are spouting up hundreds of thousands of miles above the sun's surface. It is as though the surface could not get rid of energy as rapidly as it arrived from inside, and so created a vast extra mechanism of fountains, cascades and arches of flame, to help it. These are generally of a crimson colour, and often take the most fantastic shapes. Some stand almost still as though firmly rooted in the body of the sun, but others sprout up like Jack's beanstalk, with speeds of thousands of miles a minute. Some jump clear of the sun altogether to heights of hundreds of thousands of miles, changing their shapes all the time (see Plate XIV). A prominence which starts up in the shape of a huge red mushroom may come down looking like a mangrove tree, or a fierce crimson dog, or some still more weird antediluvian animal. Plate XV shews a prominence photographed at the 1919 eclipse, which

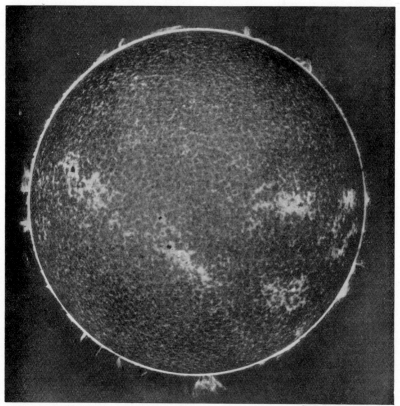

PLATE XIII

J. Evershed, Kodaikanal

The Sun photographed in calcium light

Shewing eruptions (prominences) and flocculi. See also Plate XVII (facing page 44).

PLATE XIV

8 hours 6 minutes

7 hours 52 minutes

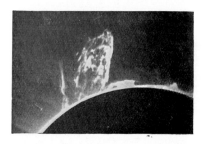

8 hours 36 minutes

8 hours 45 minutes

8 hours 57 minutes
Three stages of a prominence observed
on May 25, 1916

9 hours 3 minutes
Three stages of a prominence observed
on November 19, 1928

J. Evershed, Kodaikanal

Solar prominences of the leaping kind

That shewn in the three photographs on the right jumped to a height of 567,000 miles
above the sun's surface.

PLATE XV

A. C. Crommelin, Greenwich Observatory

The Ant-eater prominence (May 29, 1919)

The whole length of this prominence was about 350,000 miles.

PLATE XVI

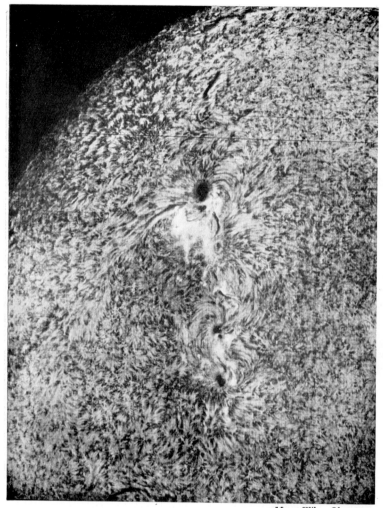

A small part of the Sun's surface photographed in hydrogen light

Shewing details of a group of four sun-spots, and the stormy formation of flocculi surrounding them.

looked for all the world like a huge ant-eater, 350,000 miles from snout to tail—the size of animal that could gulp down the whole earth like a pill. After this photograph of it had been taken, the creature raised its snout and tail up off the surface of the sun. Then it increased the number of its legs, and started to jump upwards. It jumped to a height of 475,000 miles, and then the setting of the sun made it impossible to watch its further antics.

This bizarre architecture of crimson flame is not the only scenery of the sun's surface. Here and there we see enormous dark yawning cavities, looking rather like the craters of volcanoes in eruption, belching out fire and matter from the sun's interior (see Plates XVI, XVII and XIX). On earth we call them sunspots; now that we are near them we can see that, whatever else they may be, they are not spots; many of them are so big that our whole earth could fall into them like a hard-boiled egg into a crevasse.

And now the sun fills almost the whole sky in front of us. We see it as a vivid disc of fire, which comes ever nearer and nearer; soon our rocket must crash and we brace ourselves to withstand the shock. Now the arches and fountains of flame are not only all round us; they are also above us. We are inside the sun's fiery atmosphere, so that there is light on all sides of us. If we draw a sample of this atmosphere inside our rocket and analyse it, we shall find it is of very different constitution from the earth's atmosphere. It is true it contains the same light gases as the earth's atmosphere, but it also contains heavy metallic substances such as platinum, silver and lead, and indeed most, if not all, of the substances we have on earth. These all exist in the sun's atmosphere, and all are in a vaporised form because the heat is too great for any of them to remain in either solid or liquid form. We knew all this before we left the earth, because the instrument known as the spectroscope analyses the light of the sun and tells us from what kinds of atoms it originates.

Inside the Sun

Still we wait for the crash, and wait and wait, and still it does not come. We must have travelled hundreds, thousands, now tens of thousands, of miles inside the sun, and still we meet no solid surface. Gradually we realise what has happened. We are already deep in the sun's interior—and we find nothing but gas! Even the sun's outer atmosphere was too hot for any substance to remain in the solid or liquid state. Here it is hotter still, so that naturally everything is vaporised. On the earth and moon, and probably on all of the planets, there is a sharp transition from atmosphere to solid substance: on the sun, and the stars in general, there can be no such sudden transition. The atmosphere merges gradually into the main substance of the star, which is made of the same kind of stuff as its atmosphere. As there is no solid barrier to stop the progress of our rocket, its momentum will carry us right to the centre of the sun.

Even while we were passing through the sun's weird surface scenery of fountains and arches of flame, and of twisting fiery growths, our thermometer stood at some seven or eight thousand degrees; by the time we were well inside the sun's atmosphere it had reached nine or ten thousand degrees. It was from here that we got our last glimpse of the earth through a veil of fire which was rapidly closing in around us. Then, as we dashed into the sun's interior and its fiery light entirely surrounded us on all sides, our thermometer began to run up very rapidly; it soon reached millions of degrees, and now that we are near the centre of the sun, it records something like 40 million degrees. When we get back to earth it will not be easy to form any conception of what such temperatures mean, but one single concrete fact may help our imagination. If we could take an ordinary shilling out of our pockets, and heat it up to the temperature of the sun's centre, its heat would shrivel up every living thing within thousands of miles of it.

If possible, the pressures outside our rocket are even more remarkable than the temperatures. At the earth's surface, there is an air pressure of about 15 lb. to the square inch; this is needed to support the weight of the atmosphere, and so we call it a pressure of one atmosphere. The pressure inside the boiler of a modern express locomotive is about twenty atmospheres—that is to say the steam inside the boiler exerts twenty times as much pressure as the air outside. But the pressure at the centre of the sun is about 40 thousand million atmospheres. Whereas the weight of the earth's atmosphere creates a pressure of one atmosphere at the earth's surface, the far greater weight of the whole substance of the sun creates a pressure of 40 thousand million atmospheres at the centre of the sun.

Ordinarily heating up any substance causes it to expand, while subjecting it to high pressure causes it to contract. The substance of the sun's centre wants to expand, because it is heated—to about 40 million degrees—and also wants to contract because it is subjected to pressure—to about 40 thousand million atmospheres. The tug-of-war between these two opposing influences results in compression winning, although only by a little. The substance of the sun's centre is not enormously compressed—the extreme heat prevents that—and yet, as we shall see in a moment, it is compressed more than anything we ever experience on earth.

Even the Atoms are Shattered

We saw how even the temperature of a few thousand degrees at the sun's surface was enough to turn all ordinary substances into vapour. Not only does it melt ice into water, and turn water into steam; it even loosens the joints of the separate little molecules of steam, and breaks each up into the three atoms of which it is formed—two of hydrogen and one of oxygen. We knew all this before we started on our journey, because our spectroscopes had told us that practically all the light of the sun and stars comes from molecules which have already been broken up into their component atoms: only in a few of the very coolest stars do we find a small number of unbroken molecules, and these are of specially tough kinds.

In the atmospheres of the hotter stars, our spectroscopes shew that even the atoms themselves are beginning to be broken up by the intense heat. Every atom has a very important and very massive particle known as its nucleus at its centre; ranged round this are a number of less important and less massive particles known as electrons. All the electrons are exactly similar, and so are interchangeable. But the nuclei are neither similar nor interchangeable; the nucleus of an atom of hydrogen is different in all sorts of ways from the nucleus of an atom of oxygen. Indeed it is this difference in the nuclei that causes the whole difference between hydrogen and oxygen.

This, then, is all that an atom consists of—one nucleus and a lot of electrons. All these tiny particles are charged with electricity, so that each nucleus attracts its electrons round it. It holds the two which are nearest to it in a very firm grip; a certain number more, generally eight, further out, in a less firm grip; and the remaining electrons, still further out, in a still weaker grip. Indeed the outermost electrons of all are so weakly held that even the feeble heat of a candle-flame or a coal-fire will shake some of them loose. Thus we must expect a number to break loose in the far greater heat of the atmospheres of the sun and stars. The complete oxygen atom consists of a nucleus with eight electrons round it, and our spectroscopes shew that, in the atmospheres of the hottest stars of all, many atoms of oxygen have already lost two, and some even three, of their electrons. The spectroscope cannot penetrate into the still hotter interiors of the stars, but we may be

sure that there the oxygen atoms will have lost even
more than two or three of their eight electrons. And
as we approach the centre of the sun, at a temperature of many millions of degrees, the oxygen atoms
must be completely broken up. We know the strength
of the grip by which the oxygen nucleus holds its
innermost electrons, and it is not enough to withstand the terrific heat of the sun's centre. Strictly
speaking, there are no atoms of oxygen at the centre
of the sun, only a miscellaneous collection of nuclei
and electrons dashing hither and thither in complete
disorder.

There are other types of atoms, more massive than
oxygen, in which the nucleus holds the innermost
electrons in so firm a grip that even a temperature of
40 million degrees fails to loosen them. Thus even
at the centre of the sun, some of the nuclei of these
atoms must still have their two innermost electrons
attached, forming a sort of very miniature atom.
The substance of the sun's centre will consist of
innumerable shoals of these miniature atoms, with
the still smaller detached electrons, as well as the
completely broken fragments of other atoms, flying
about helter-skelter through and between them.

All these objects are moving with terrifically high
speeds, the result mainly of the extreme heat. If we
could measure the speed of the broken-off electrons
as they flash past the windows of our rocket, we
should find that they averaged about 30,000 miles a
second—some 100,000 times the speed of an ordinary
rifle-bullet. We can see clearly enough that the
broken fragments of the atoms cannot re-form into

complete atoms while they are under continual bombardment by projectiles moving at such speeds as this.

A Tour in Time

Before we steer our rocket back to earth, let us call on it for one more service which it is quite capable of rendering; let us call on it to take us backwards in time.

Let us go back some 3000 million years in time and then cruise about in space somewhere near to the sun, and watch the years roll by. Strictly speaking, there are no years yet. For a year is the time the earth takes to go completely round the sun, and at the time at which we now are, there is no earth. We have gone back not only to a time before man trod the earth, but to a time before there was any earth for him to tread.

Yet we notice that the sun looks much the same as it does to-day; it is very slightly larger, slightly more luminous, slightly hotter. The 3000 million years which we have stepped back in time form but a day in the life of the sun; it has hardly aged perceptibly in the interval.

On the other hand the sky is quite unrecognisable to our eyes of 1931 A.D. The stars do not travel a great distance in the span of a man's life, but they travel so far in 3000 million years that we cannot recognise any of the familiar landmarks or constellations. The sky looks as foreign to us as the southern

sky does to a traveller who has just arrived from the
north.

As the years roll by in their thousands and millions,
the appearance of the sky continually changes. Constellations change their shapes, and the stars their
brightness, as they approach and recede. A star
which at one epoch was the brightest in the sky
recedes until it becomes quite faint, and finally dis-
appears from sight. We notice that there is seldom a
star in the whole sky which looks as bright as Sirius
does to-day; we begin to realise that Sirius presents
a combination of nearness with intrinsic brightness
which is rather rare. Yet on one occasion at least
Sirius was completely outdone in brilliancy.

Our World is Born

As we cruise about near the sun, and watch the
changing panorama of the sky somewhere between
two and three thousand million years ago, we notice
a star gradually increasing in brightness until it out-
shines all the others in brilliancy, and finally looks
incomparably brighter than Sirius does now. It
looks bright because it is very near rather than because it is intrinsically very bright; indeed, it has
approached quite unusually near to the sun. And
as we watch, it comes ever nearer; it is heading
almost straight for the sun. It is no longer a mere
point of light. We see it as a large disc. And now it
has come so near that its mechanical effects are beginning to shew. Just as the moon, by its nearness to
the earth, raises tides in our oceans, so this enor-

PLATE XVII

J. Evershed, Kodaikanal

The Sun photographed in hydrogen light

Shewing sun-spots and also bright filaments of hydrogen gas.

PLATE XVIII

N.G.C. 5278–9

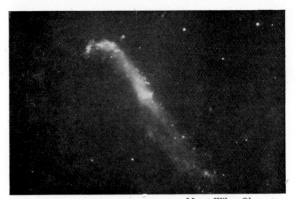

Mount Wilson Observatory

N.G.C. 4656–7

Tidal Action

Although the interpretation of both these photographs is doubtful,
they may serve to illustrate how a filament may be pulled out of a
gaseous mass by the tidal action of a second body.

mously more massive body is, by its nearness, raising tides in the fiery atmosphere of the sun. Because it is so much more massive than the moon, these tides are incomparably greater than those which the moon raises in the earth. They become so great that, at a point right under the star, the sun's atmosphere is drawn up to form a huge mountain, many thousands of miles high. This mountain travels over the surface of the sun, keeping always under the star which causes it, as this moves on its way through space. At the opposite point of the sun's surface, another but smaller mountain keeps always opposite the main one. As the star approaches ever nearer, these tidal mountains continue to increase in height, until at last, when the other star is so near as to fill up a large part of the sky, a new feature enters. So far, the gravitational pull of the star has been drawing up the summit of the larger mountain in opposition to the gravitational pull of the sun, but the latter has always been the stronger. Now the second star comes so near that the balance suddenly swings over in the other direction; the second star outdoes the sun in gravitational pull, and the top of the mountain shoots off towards it. As this relieves the pressure on the lower parts of the mountain, these also shoot upwards, and then the parts below them, and so on, so that a whole stream of matter shoots out from the sun towards the second star. If this star came continually nearer to the sun, the end of the jet of matter would reach it in time, and the substance of the jet would join the two stars together like the bar of a dumb-bell (see Plate XVIII).

Actually the other star is not heading directly towards the sun; after coming very near indeed, it finally passes on its way without actually colliding. As it recedes its tidal pull diminishes. No more matter is pulled off the sun, and the jet which has already come off forms a long filament of hot filmy gas suspended in space. In shape, it is rather like a cigar, pointed at its two ends. The point which is now furthest from the sun was originally the peak of the tidal mountain. The thick middle of the cigar consists of the matter which came off plentifully when the star was nearest, and its tidal pull was strongest. Finally the pointed end nearest the sun is formed of the last thin dribble of matter which came off just before the tidal pull became too weak to draw any more matter away from the sun.

Even as we watch this cigar-shaped filament of fiery spray, it gradually cools and, as it does so, it condenses into detached separate drops, much as a cloud of steam condenses into drops of water. Yet these drops, like the filament itself, are colossal structures; their size is on the astronomical scale. Naturally they are biggest near the fat centre of the cigar, where the matter of the filament was most abundant, and are smallest at the two ends.

Finally, these detached drops of matter begin to move about in space as separate bodies. They do not fall back into the sun, because the pull of the other star, which we now see receding in the distance, has set them in motion; unless they happen to be moving directly towards the sun, they will not fall into it but describe orbits round it. This is a direct consequence of the law of gravitation, which was the same thousands of millions of years ago as it is to-day. Some of these orbits may be nearly circular while others are greatly elongated. As we watch the orbits for millions upon millions of years, we see them gradually and very slowly changing their shapes. The condensed drops of matter do not move in un-obstructed paths, for the great cataclysm we have just witnessed has left space littered with its debris. The great drops must plough their way through this, and as they do so the shapes of their orbits gradually change, until at last, after thousands of millions of years, they move round the sun in almost circular orbits, just like the planets of to-day. And indeed these bodies are the planets; the dramatic spectacle we have just witnessed from our imaginary rocket is one which must inevitably happen in Nature whenever one star approaches close enough to another, and its final scene is so exactly like the solar system, that we have every reason to suppose that this is actually the way in which the planets came into being. So far as we can judge from their present arrangement and movements, it seems most likely that they were torn off the surface of the sun by the tidal pull of a passing star which happened to pass very unusually near to it some few thousands of millions of years ago.

We have already noticed how the sun's atmo-sphere contained platinum, lead, and most of the substances that we find on earth. We now see that it must inevitably contain precisely the same substances as the earth, since the earth is nothing but a sort of solidified sample of it. We cannot of course tell what other substances there may be down in the far depths of the sun, since these have no means of revealing themselves to us. But it is significant that practically all the substances which occur on earth are observed spectroscopically in the sun's atmosphere, and so far no reason has been found for thinking that this contains any substances which do not occur on earth.

CHAPTER III

THE SUN'S FAMILY

Now that our rocket has brought us safely back to the earth of to-day, let us consider in more detail the small colony, almost completely isolated in space, which we believe to be the shattered fragments of what was once an ordinary star. It contains a great variety of objects, large, medium-sized, small and very small, which we must discuss in turn.

The Nine Planets

First let us look at its largest members, the nine principal planets. These move round the sun in almost circular paths, rather like circus horses trotting or galloping round the ring-master. They all go round in the same direction, and this must of course be the direction in which the wandering star, which brought them into being, moved round the sun. Because of the way it came into existence, the solar system has only one-way traffic—like Piccadilly Circus. The traffic nearest the centre moves fastest; that further out more slowly, while that at the extreme edge merely crawls— at least by comparison with the fast traffic near the centre. It is true that even the furthest and slowest of the planets covers nearly three miles every second, which is about 200 times the speed of an express train, but this is a mere crawl in astronomy. The

planets Mercury and Venus, which constitute the
fast traffic near the centre, move, the former ten and
the latter seven, times as fast. We shall find the
reason for all this later; at present we are merely
concerned with the facts.

Before we leave Piccadilly Circus, it should be un-
derstood that we cannot represent the solar system
by putting up a statue of Eros in the middle to represent the sun, and letting nine taxicabs gyrate round it
to represent the nine planets. The statue is far too big
to represent the sun, and the taxicabs are enormously
too big to represent planets. If we want to make a
model to scale, we must take a very tiny object, such
as a pea, to represent the sun. On the same scale the
nine planets will be small seeds, grains of sand and
specks of dust. Even so, Piccadilly Circus is only just
big enough to contain the orbit of Pluto, the outermost
planet of all. Think of a pea and nine tiny seeds, grains
of sand and specks of dust in Piccadilly Circus, and
we see that the solar system consists mainly of empty
space. It is easy to understand why the planets look
such tiny objects in the sky.

Yet the solar system is very crowded compared
with most of space. If a pea and nine smaller
objects in Piccadilly Circus represent the sun and
planets, the nearest of the stars will be represented
by a small seed somewhere near Birmingham—all in
between is empty space. Again we see how isolated
the solar system is in space.

PLATE XIX

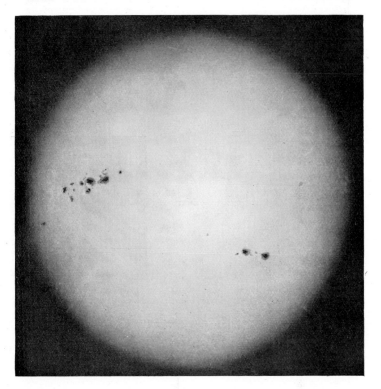

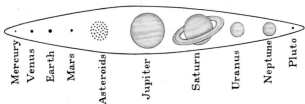

Mercury | Venus | Earth | Mars | Asteroids | Jupiter | Saturn | Uranus | Neptune | Pluto

The Sun and Planets drawn to scale

The planets are arranged in order from the sun, and we see how their size increases until Jupiter is reached and then diminishes again.

Only the sizes, not the distances, are drawn to scale. If the distances were to scale the earth would be 11 yards, and Pluto a quarter of a mile, away from the sun.

PLATE XX

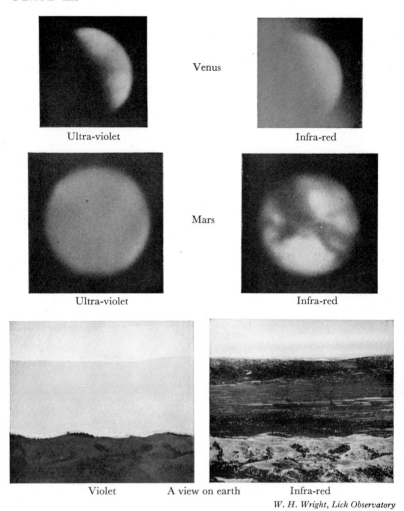

Ultra-violet   Venus   Infra-red

Ultra-violet   Mars   Infra-red

Violet   A view on earth   Infra-red

W. H. Wright, Lick Observatory

Venus and Mars

The two bottom pictures shew how violet light photographs mainly the atmosphere, the distant town (San Jose) hardly shewing through $13\frac{1}{2}$ miles of air, while infra-red light penetrates $13\frac{1}{2}$ miles of atmosphere easily. Venus is so surrounded by atmosphere that even infra-red light fails to find a solid surface. Mars has a distinct atmosphere, as is shewn by the ultra-violet photograph on the left, but infra-red light penetrates this easily and photographs the permanent surface of the planet. Careful measurement shews that the ultra-violet image is substantially larger than the infra-red, indicating that the Martian atmosphere extends to a considerable height.

Mercury

Now let us consider the planets in detail. Nearest to the sun comes Mercury. It is so near that we always see it near to the sun in the sky. The Greeks had a legend that Mercury was the close and intimate friend of Apollo, the sun. They are unfortunately such inseparable friends that we never see Mercury in the night sky; this would take it too far away from the sun. If we have no telescope, the best we can hope is to see it in the western twilight as an evening star, just after the sun has set, or else in the east, as a morning star, just before the dawn. Even this calls for good luck, and unfortunately nine days out of ten the luck is not good in our latitudes: Mercury generally lies hidden in the clouds or mists of the horizon. It is seen far more easily in lower latitudes.

As Mercury travels round the sun, it is sometimes on the side of the sun nearest us, and sometimes on the further side. When it comes exactly between us and the sun, the light of the sun falls only on the side away from us, so that the face it turns to us is all in darkness. On these occasions Mercury can be seen as a small black disc, crossing in front of the bright disc of the sun. When it is in other positions, we can see part of its illuminated face from the earth. The part we see illuminated may range through all shapes from a thin crescent, like a new moon, to the full circle we see when the planet is at the far side of the sun. Because of this, Mercury appears to go through phases, like the phases of the moon. The unilluminated part of its face is always entirely dark, and this shews that it does not shine by its own light, but only in virtue of the sunlight which falls on it. And the same is found to be true of all the planets.

Venus

Venus, which comes next in order, is nearly twice as far from the sun as Mercury, but is still so near that it is only rarely seen in the night sky. Like Mercury, it is generally seen either in the twilight as an evening star or in the light of the dawn as a morning star. After the sun and moon, it is far and away the most brilliant object in the sky.

Venus exhibits phases like Mercury and the moon, the result of our usually not seeing the whole of its illuminated hemisphere. Also, as it moves round the sun, its distance from us changes so much that it seems to vary almost as much in size as in shape.

It looks largest when it is at its nearest, almost exactly between us and the sun. Its apparent shape is then that of a fine crescent, like the new moon. All the rest of the face it turns towards us is in darkness. When it is furthest from us—almost exactly behind the sun—it is nearly six times as distant, and so looks only a sixth as big as it does when at its nearest. On these occasions, the sun's light falls on the whole of the face it turns towards us, so that it looks circular in shape, like the full moon.

Its apparent brightness changes with its shape and distance, and it looks brightest when it has the crescent shape of a five-day-old moon (cf. Plate IX). It then appears twelve times as bright as Sirius, and would be terrifically dazzling, were it not that its nearness to the sun prevents its being seen to full advantage. Yet when the sun's brightness dims the light of Venus, it dims that of other and fainter stars still more, so that as evening closes in Venus is often the first star to appear in the deepening twilight of the western sky. At other times Venus may form a particularly brilliant "morning star," often being the last star to fade away in the light of day. For this reason, it is commonly supposed to have been the "star of Bethlehem" which the Magi saw in the East. At times it may be so bright that even the full light of the sun cannot entirely obliterate it. It has frequently been seen in full daylight, sometimes even at noon, by the unaided eye. With a telescope of even moderate power, we can follow its motion as it journeys across the sky by the side of the sun in broad daylight, from morning until evening.

The Earth

Next after Mercury and Venus in order of distance from the sun comes our earth. It is larger in size than either, although only very slightly larger than Venus. The order of increasing size, Mercury, Venus, the earth, is that of increasing distance from the sun, which accords well with the supposition that the planets were formed as condensations out of a cigar-shaped filament of gas. Mercury, the smallest planet of all, would of course be the pointed end of the cigar. (See Plate XIX, p. 50.)

We have seen how Mercury and the moon, both much smaller than the earth, are devoid of atmosphere, their gravitational pulls being too feeble to retain one. Venus and the earth are both large enough not to suffer from this disability.

As Venus and the earth are of about the same size, and have in all probability had similar life-histories, we might reasonably have expected to find that their atmospheres would be similar. Actually they are very different. In particular, oxygen, which forms a large fraction of the earth's atmosphere, appears to be exceedingly rare, if it exists at all, on Venus. We know that oxygen combines very freely with other substances; such combination occurs, for instance, when substances burn or corrode or rust. This being so, we need not feel surprised that there is little or no oxygen left in the atmosphere of Venus; what would be surprising, if we did not know the explanation, is that there is so much left in the atmosphere of the earth. The explanation is that every tree and every blade of grass on earth is a sort of oxygen factory; the earth's vegetation keeps up the supply of oxygen. The circumstance that no appreciable amount of oxygen can be detected in the atmosphere of Venus goes far to suggest that there is no vegetation on Venus, and so probably no life of any kind.

The Outer Planets

Mercury and Venus are so near to the sun in space that we always see them near to the sun in the sky. The six planets which we have not yet discussed move round the sun in circles which lie beyond the orbit of the earth. As we look out at these other planets from our position near the sun, they appear to journey not only round and round the sun, but also round and round us, so that we frequently see them in directions right away from the sun in the dark night sky. Under these conditions, the two planets which lie nearest the earth, Mars and Jupiter, can both form imposing objects—indeed, at their best either of them may be the brightest object in the whole of the sky. They are only about a tenth as bright as Venus, but Venus is a lamp which burns mainly in daylight or twilight, while Mars and Jupiter are candles which burn in the darkness of the night; they do not have to compete with the fierce light of the sun. All the other planets are far fainter than these. Saturn, the brightest of them, only looks like a very ordinary star. Uranus can just be seen by the unaided eye, but Neptune is well below the limits of visibility, and Pluto is enormously below; we need quite a powerful telescope to see Pluto.

Mars, which comes first as we recede outwards into space, is much smaller than the earth, its diameter being only a little more than half of the diameter of the earth. Thus Mars temporarily breaks the rule that the planets increase in size as we recede from the sun. But Jupiter, which comes next, restores it in full force. It has nearly eleven times the earth's diameter, and 317 times its weight; indeed it has more than double the weight of all the other eight planets rolled into one. As it is the middle one of the series of planets, fifth out of nine, it must have been born out of the middle part of the cigar-shaped filament, where the matter was richest. This fits in with its being the largest and most massive of all the planets. Beyond Jupiter, both the sizes and weights of the planets steadily decline; we have passed the middle of the cigar, and are approaching its thin end. Saturn, which comes next in order, contains less than a third of the substance of Jupiter, while the other three planets are enormously smaller than Saturn. Indeed Pluto, at the other pointed end of the cigar, appears to be hardly larger than Mercury.

The Climates of the Planets

A telescope is primarily an instrument for collecting a great amount of light from a star, or a group of stars, and sending it all into a man's eye or onto a photographic plate—just as an ear-trumpet collects a great amount of sound, and sends it into a deaf man's ear. A telescope also collects a great deal of heat, and instruments have been devised for measuring this heat with extreme accuracy. They are now made so sensitive that a big telescope could measure the heat sent out by a single candle hundreds of miles away; it quite easily measures the heat sent out by the nearer planets and the brighter stars.

Broadly speaking it is found that the planets send out just about as much heat as they receive from the sun, and no more. We have long known that they shine only by reflected light—in other words that the light they emit is merely the light they receive from the sun and reflect back—and now the same proves to be true of their heat. When they first came into being as flecks of fiery spray thrown off from the sun, they must have been intensely hot, and must have sent out fierce heat of their own, but 2000 million years have intervened since then, and have given them ample time to cool down. They no longer have any heat of their own, and are warm only in so far as they are warmed by the sun. As a consequence, the further they are from the sun, the colder they are —like campers sitting round a camp-fire.

Indeed, we may think of the sun and stars as vast collection of camp-fires scattered through space Out in the remote depths of space, far away from these fires, the cold is intense—about 480 degrees of frost. As we move inwards towards the sun, or of course any other camp-fire, we come to more genial temperatures, but we have to move in a long way before we reach a state of things which can be fairly described as comfortable—or, indeed, on which life can be possible at all. The outer planets, Pluto, Neptune, Uranus and Saturn, must be colder than anything we ever experience on earth. Even Jupiter is almost unimaginably cold. The amount of heat we receive from it shews that its temperature must be about 270 degrees below zero on the Fahrenheit scale. This is so cold that not only would water be frozen, but the commonest gases, like those of our own atmosphere, would be turned into liquids. Yet the planet is not altogether devoid of activity; definite features appear in its atmosphere which persist for a time, and then disappear much as rain-clouds do in the atmosphere of the earth (see Plate XXII, p. 63). The clouds of Jupiter must presumably be clouds of carbon-dioxide, or of some other gas which only condenses at very low temperatures.

When we come to Mars, the planet next to the earth, we encounter less inhospitable conditions, yet the surface even of Mars is mainly below the freezing-point. A spot on its equator at noon, with the sun beating right down from immediately overhead, is probably about as warm as London on a November afternoon. But, as we know, Mars has very little of an atmosphere to retain this heat. Also the quality of the light it sends us suggests that its surface, like that of the moon, consists mainly of volcanic ash, which again has no power of storing heat. Thus, as the sun moves on and night approaches, the temperature falls very rapidly; frost sets in before evening, and by midnight it must be about as cold on the equator of Mars as it is at our North Pole.

Our earth is at what may be described as a comfortable temperature, but as we pass on, and move still nearer to the sun, we find that the two inner planets, Venus and Mercury, are not, Venus being distinctly too warm for comfort, and Mercury very much so. A "place in the sun" on Mercury is just about as warm as a place on a grill over a hot fire.

Is there Life on Mars?

Thus the earth is the only planet whose temperature seems entirely suited to life of the kind we know. Its most serious competitor is its chilly neighbour Mars. Many astronomers have seen markings on Mars, which they describe as canals, and believe to have been artificially constructed. On the other hand, photographs of the planet give no indication that intelligent beings have left their marks on its surface. The evidence for the existence of such marks has so far rested almost entirely on direct visual observation, and the human eye is notoriously capricious and unreliable when it is forced to work with inadequate light. Various experiments have proved that the eye which is struggling to study outlines in faint light tends to connect up the light and dark patches on a dimly lighted object by non-existent straight lines, like those which the early observers of Mars believed they saw on the planet. It is in keeping with this that early observers claimed to see very similar markings on Mercury and Venus. Yet we now know that the visible surface of Venus consists only of clouds, and Mercury is obviously unsuited for life. At a still earlier period, observers used to put markings of similar type on their charts of the moon; some of these are known to be wholly imaginary, while others, which really exist, are certainly not canals. The general life-history of such markings appears to be that they are originally put into drawings made with inadequate light and inadequate magnifying power, and subsequently disappear in the light of fuller knowledge. In view of all this, most scientists will suspend judgment on the supposed evidences of life on Mars until the camera confirms that they really exist.

There is, however, no question that certain seasonal changes are observed on Mars. During the planet's winter a white snow cap undoubtedly forms round the North Pole. In summer this melts and, as it does so, the country further south changes its appearance. Some astronomers think these changes may be caused by the growth of vegetation, promoted by the flow of melted ice; to others it seems more likely that they are caused by rain falling and watering a lifeless desert of volcanic ash.

On the whole, the case for life existing on Mars, or on any other planet in the solar system, can hardly be called a strong one, and, although there is still room for much difference of opinion, it seems to me most likely that the life which exists on our earth is the only life in the sun's family, although other stars far out in space may include inhabited planets in their families.

The Satellites of the Planets

Most of the planets are accompanied by retinues of satellites, or moons, proportional in number to the size and dignity of the planet. The two largest planets, Jupiter and Saturn, each have nine; Uranus, which comes next in size, has four, while the still smaller planets have two, one, or even none at all.

We believe the satellites to be fragments which were torn out of the planets, just as the planets are fragments torn out of the sun, and by much the same sequence of events.

For mathematical theory shews that there is what may be a danger-zone surrounding every big body in space. A small body enters this danger-zone as soon as it comes within a certain calculable distance of the big body. And, when it does so, the gravitational pull of the large body proves too much for it and tears it to pieces. No small body can enter the danger-zone of a big body and come out entirely whole, although the amount of damage done will depend on the length of stay inside the danger-zone. We believe that, long ago, in its blind wanderings through space, the sun entered the danger-zone of a bigger and more massive star, and paid the usual penalty of being broken up in the way already described (see p. 45, above). Matter was torn out of it, forming a cigar-shaped filament, out of which the planets were born. We have seen how these planets would not at first describe the regular circular orbits in which they now move; their motions would be far more erratic, and might take them into the danger-zone surrounding the sun, in which case they would be broken up much as their parent, the sun, had been broken up before them. It seems most likely that the satellites of the planets were born in this way. Indeed, the satellite systems are so much like tiny replicas of the main solar system that we are almost compelled to suppose that they were produced by the same process as the main system. If so, the sun is the parent of the planets, and the grandparent of their satellites.

Saturn's Rings

Saturn is in many ways the most interesting of all the planets, and is certainly the most sensational in appearance. It not only has nine moons, but also is surrounded by three flat circular rings, which form a sort of frill or collar round its middle (see Plate XXI). Galileo first discovered these in 1610, and many speculations were made as to their true nature. In 1750 Thomas Wright made the suggestion that, "Could we view Saturn through a telescope capable of it, we should find his rings no other than an infinite number of lesser planets, inferior to those we call his satellites."

The truth of this conjecture has been proved up to the hilt. In 1859, Maxwell, the Cambridge mathematician, who described the rings as "from a purely scientific view, the most remarkable bodies in the heavens," proved mathematically that they must be of the nature which Wright had conjectured. In 1895 the American astronomer Keeler clinched the matter with some observations which shewed that the substance of the rings was moving round and round the planet, but that the outer parts of the rings moved more slowly than the inner. Again, as in the solar system as a whole, we find one-way traffic with the slower traffic keeping to the outside. We could not possibly find this if the rings were solid, but we

PLATE XXI

1916

1921

1909

1912

Lowell Observatory

Saturn

Photographs taken in four different years shewing four different aspects
of Saturn and its system of rings.

PLATE XXII

Ultra-violet

Violet

Red

Infra-red

W. H. Wright, Lick Observatory

Jupiter

The planet is photographed in four colours of light. Ultra-violet and violet
shew most detail, suggesting that the markings on Jupiter are atmospheric,
as is also shewn by their non-permanence.

could find nothing else if they consisted of millions of miniature moons.

There is every reason for thinking that these tiny moons are the fragments of a body which at one time formed a quite ordinary full-sized moon of Saturn. This probably entered the danger-zone of Saturn, which no small body can enter with impunity, and paid the usual penalty of being shattered into fragments. Just as we believe that in the remote past a passing star broke up the sun, and so formed its present family, or again as we believe that the sun broke up Saturn and formed its satellites, so we believe that Saturn itself broke up its nearest satellite into millions of fragments and so formed its own system of rings—a third generation of astronomical bodies.

Yet the two processes were not entirely similar. The sun only made a temporary stay inside the danger-zone of the bigger star. It was moving through space at a good speed, and its motion carried it out of the danger-zone before it was completely broken up. So also, the stay of Saturn inside the danger-zone of the sun was only temporary. The satellite of Saturn, on the other hand, was describing a circular path round Saturn; the reason it got inside the danger-zone of Saturn was that this circle was shrinking gradually in size. In this way it had the bad luck to get inside the danger-zone without ever being able to get out again, and so was broken into fragments. There can be little doubt that this conjecture is sound, for we can calculate the distance to which the danger-zone of Saturn extends. Saturn's innermost satellite is just outside it, as of course it must be to remain whole, *but the rings are inside.*

Nowhere in the solar system do we find a satellite of reasonable size revolving inside the danger-zone of its planet. Jupiter's innermost satellite comes nearest, being very near indeed to the danger-zone of Jupiter. It seems likely that in course of time this satellite must draw ever closer in towards Jupiter, and that sometime in the not very remote future it must enter the danger-zone of the great planet and be broken up. Jupiter will then be surrounded by rings as Saturn now is.

In the same way, although only in the very far future, our own moon must inevitably be drawn in closer and closer to the earth, until finally it approaches too near for safety and meets the same fate. After this the earth will have no moon, but will be surrounded by a frill of rings like Saturn. These rings will not only reflect far more of the sun's light than our present moon does, but there will be full moonlight all night long. While this will no doubt add to the amenities of life, things will, in some respects, be less comfortable than now. For at frequent intervals, moons must crash into one another and their broken fragments fall onto the earth like huge rocks falling from the sky.

The Asteroids

In between Mars and Jupiter, thousands of small bodies, called "asteroids" or "minor planets," travel round and round the sun with the usual one-way

PLATE XXIII

Lowell Observatory

Halley's Comet as observed on May 7, 1910

The most famous of all comets. It was probably this comet that "hung over Jerusalem
in the figure of a sword" before the destruction of the city in 66 A.D., and appeared
just before the Conquest of England in 1066 A.D. Its appearance in 1759, confirming
Halley's predictions from the law of gravitation, convinced men that its comings and
goings were ordered by this law and not by the approach of disasters on earth.

PLATE XXIV

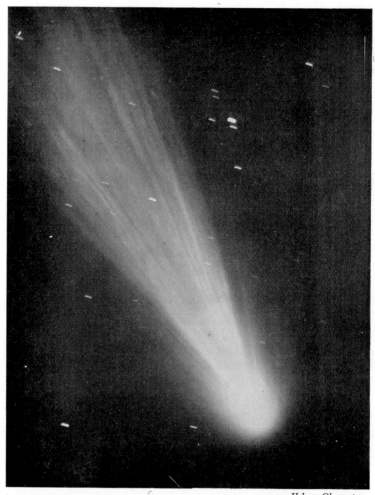

Helwan Observatory

Brooke's Comet as observed on November 3, 1911

The slight elongation of the star-trails results from the telescope having been made to follow the comet and not the stars throughout a 15 minutes exposure. (See also Plate XXIII.)

traffic of the solar system. These, too, are probably the shattered fragments of a single big body. There is an abnormally wide gap between the paths of Mars and Jupiter, and it seems likely that a single quite ordinary planet used to revolve in this space, and met its fate by entering the danger-zone of Jupiter.

Comets and Shooting-stars

The remainder of the sun's family are quite small objects. Foremost in size and importance come the comets. These resemble the planets in travelling round and round the sun, but differ from them in mostly having very elongated paths, so that at one time a comet may be far out in the cold depths of space, and at another quite close to the sun. Comets are usually invisible until they get well into the light and heat of the sun. Then they are apt to become conspicuous, and even sensational, out of all proportion to their true importance. They too are broken up when they enter the danger-zone surrounding a big body such as the sun or Jupiter; the broken fragments then form showers of stones which we call meteors. Occasionally the earth happens to pass right through one of these showers, so that some of the meteors get entangled in the earth's atmosphere. These are raised to a white heat by the friction of the air, and we have what is known as a meteoric display—a shower of shooting-stars. In a few cases the paths of these meteor showers coincide exactly with the former paths of vanished comets, giving a very convincing proof that the comets have been broken up into a swarm of smaller bodies. And, indeed, the whole history of the solar system is in large part one long story of big bodies being broken into smaller ones, not so much by direct collision as by gravitational forces, such as raise tides on our earth, tearing them to pieces.

Most meteors are no bigger than a walnut or a pea, if as big. Generally they are small enough to be completely vaporised before they strike the earth, leaving only a bright trail of luminous dust. The end of this trail marks the spot at which they became completely dissolved into vapour, and it is usually many miles above the ground. Occasionally, however, a meteor is too large to be entirely vaporised in its rapid flight through the air, and what is left of it strikes the earth as a meteoric stone. All parts of the earth are of course liable to bombardment by these stones, which appear to fall out of the skies. The Book of Joshua tells how "The Lord cast down great stones from heaven." Many other falls of stones are mentioned by early writers, and great numbers of fallen meteorites have been preserved, some of them of considerable size and weight.

In Arizona there is an enormous hole, shaped like the crater of a volcano. This is believed to have been formed by the impact of a huge meteorite, as big as a mountain, in prehistoric times. No meteorite of comparable size has fallen in recent years, although a very large one fell in Siberia in 1908, making a wind in its fall which devastated the forests for miles around; hardly a tree was left standing in the space of 100 square miles.

PLATE XXV

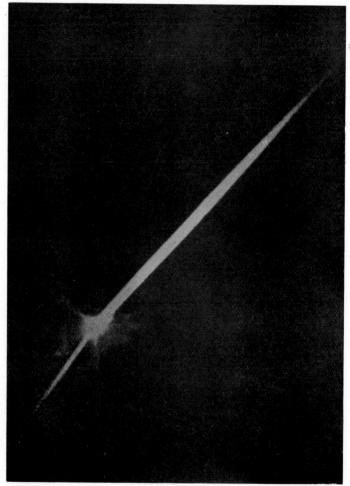

C. P. Butler

Exploding Meteor

This large meteor exploded in the air while its trail was being photographed.

PLATE XXVI

A. C. Crommelin, Greenwich Observatory

The Solar Corona at the eclipse of 1919

This was taken on the same occasion as Plate XV, but with a longer exposure.
The ant-eater prominence is faintly visible.

How old is the Earth?

Certain substances which occur both in the earth and in these meteoric stones change their composition gradually with the passage of time. By noticing how far this change has proceeded, it is possible to tell the age both of the earth itself and of the stones which fall onto it from outer space. Both the earth and the meteorites are found to have existed for about 2000 millions of years since they solidified, which seems to shew that both are the result of a cataclysmic event which occurred about this number of years ago.

Even smaller bodies circulate round the sun, like planets of infinitesimal size. These include small particles and specks of dust, single atoms and even broken bits of atoms. After the sun has set, some of these particles may reflect its light, and produce the phenomenon known as the zodiacal light. Others may reflect its light when it is hidden behind the moon at an eclipse, and produce what is called the solar corona—a sort of atmosphere of dust lighted up by the rays of the hidden sun (see Plate XXVI).

Every one of these bodies, from the great sun itself and the giant planet Jupiter down to the tiniest speck of dust in the sun's family, has its motion mapped out for it, and is controlled, by the force of gravitation which we must now discuss.

CHAPTER IV

WEIGHING AND MEASURING THE STARS

We have already seen how important the force of gravitation is, both to astronomy and ourselves. It keeps the moon tied to the earth, and maps out the paths of all the planets and other members of the sun's family; it raises the tides in our oceans, and, we believe, raised those far greater tides in the sun which, some 2000 million years ago, brought our earth, and so ultimately ourselves, into being. Finally, it keeps us alive, by making the earth stay near the sun instead of running away into the icy depths of space.

Let us try to understand a little more as to what this force is.

The Force of Gravitation

No man can lift a ton weight; he is prevented by the force of gravitation—or gravity, as we usually call it when it acts on earth. This pulls the weight to the ground, and proves too strong for him.

Again, we find it impossible to throw a cricket ball for a mile; we are prevented by the same force, which continually pulls the ball towards the ground, and invariably succeeds in getting it down before it has travelled a mile. We can easily throw the ball out of our hands at twenty miles an hour, and if gravity did not draw it earthward, it would cover a mile every three minutes, and after a year it would be far out in space, 175,000 miles away from the earth. Actually gravity interferes with this motion by continually pulling the ball earthward.

Or again, to take an example of another kind, the moon is at present moving over our heads at about 2300 miles an hour; if it were not pulled earthwards, it would continue to move in the same direction as now with the same speed as now, and a year's travel would take it far away from the earth, twenty million miles out in space. Instead of this, it moves round and round the earth; its path continually curves towards the earth, like the path of the cricket ball.

Sir Isaac Newton saw that this continual earthward curving of the moon's path could only mean that the earth was continually exerting a pull on the moon. He conjectured that this pull must be similar to that which the earth exerts on objects near its surface. Tradition says this was first suggested to him by the fall of an apple in his garden. This led him to propound his famous law of gravitation, according to which every object in the universe exerts a pull on every other object, no matter how distant it may be.

Of recent years Einstein has shewn that Newton's mathematical statement of the law was not absolutely accurate. Also the pull has proved to be of a rather different nature from that conjectured by Newton; we no longer think of it as a mere mechanical force, like the pull of a locomotive on a train. But for our present purpose the difference between Newton and Einstein is unimportant.

Studying Gravitation

Scientists can study this gravitational pull in detail, both in terrestrial laboratories and in that far greater laboratory of the skies in which Nature for ever performs experiments, on her own colossal scale, and allows us to watch the results.

The more massive a body is, the greater its gravitational pull is found to be. The great bulk of the earth is so incomparably more massive than anything else we encounter in our ordinary lives that we are generally unconscious of any gravitational pull except that of the earth, and so are tempted to think of gravitation as inherent in the earth alone. Yet delicate measurements, such as can only be made in the laboratory, shew that every object exerts its own gravitational pull.

As with all forces between bodies, the gravitational pull of a first body on a second is exactly equal in amount to the gravitational pull of the second body on the first. For this reason, it is perfectly correct to speak of the gravitational pull *between* two bodies A and B. This means either the pull of A on B, or the pull of B on A; they are precisely the same in amount. The fall of an apple to the ground provides direct evidence of the pull of the earth on the apple, but it is less easy to obtain evidence of the exactly equal pull of the apple on the earth. A pull of this amount has a large effect on a small body such as an apple, but its effect on the huge mass of the earth is quite inappreciable.

The gravitational pull between two objects is found to depend on the amounts of substance they contain, but not on the nature of their substance. For instance, the earth exerts the same pull on a ton of lead as it does on a ton of water, or on a ton of sand, or on a ton of any other substance. This is the scientific principle which underlies all ordinary commercial weighing. When the grocer weighs out a pound of tea, he is, in effect, balancing the earth's pull on a quantity of tea against the earth's pull on a standard quantity of iron or brass in the form of a pound weight. If the pulls are the same, the amount of tea is equal to the amount of iron or brass in the weight.

Two tons of substance exert exactly twice the gravitational pull of one ton, and so on. This is why the grocer can weigh out two pounds of tea by balancing its pull against the combined pulls of two separate pound weights.

Weighing the Earth

If, however, we move two objects further apart, the gravitational pull between them is found to decrease. We know exactly how the pull decreases with increasing distance, so that we can always allow for the effects of distance. The experimenter in the laboratory can measure the pull which a ton of lead exerts on another ton of lead at a known distance away. Knowing this, we can reckon what weight the earth must have to exert its gravitational pull on a ton of lead or on a flying cricket ball or on the moon. The gravitational pull of the earth, whether exerted on a ton weight or on a flying cricket ball or on the moon, shews that its weight is just about 6,000,000,000,000,000,000,000,000 tons.

Weighing the Sun

Since Newton's day, the facts of astronomy have proved, beyond all reasonable doubt, the truth of his conjecture that this force of gravitation acts throughout the whole of space; every body pulls every other body towards it, no matter how distant it may be. Newton's apple not only exerted its pull on the earth, but on every star in the sky, and the motion of every star was affected by its fall. We cannot move a finger without disturbing all the stars.

It is through this force of gravitation that the sun controls the motions of all its vast family of planets, comets, meteors and so on, from the huge bulk of Jupiter down to the smallest speck of dust which makes its humble contribution to the zodiacal light or the solar corona. We know this because they follow the paths which can be predicted for them from the law of gravitation.

Just as we can calculate the earth's weight from its pull on the moon, so we can calculate the sun's weight from the pull which it exerts on the earth or on any other planet to keep it from running off into space. All the planets agree in assuring us that the sun has 332,000 times the weight of the earth; for every ounce of substance in the earth, the sun has very nearly a ton.

Because the sun has this huge weight, its gravitational pull is tremendous. On the surface of the sun, a strong man would hardly be able to lift a 7 lb. weight, and would not be able to throw a cricket ball for more than two or three yards. He could not even perform these modest feats unless he was made of steel; a man of ordinary flesh and blood would simply be crushed flat under his own weight.

While the sun exerts this huge gravitational pull on all the members of its family, these also exert their own smaller pulls on one another. For instance, any planet, asteroid or comet whose path takes it near to Jupiter, is pulled conspicuously out of its course by the gravitational pull of the giant planet. Indeed it has been suggested that the two outermost moons of Jupiter may not have been born out of Jupiter at all, but may be asteroids which the huge gravitational pull of Jupiter has "captured" for it —i.e. pulled so far out of their original orbits that they have been compelled to revolve round Jupiter ever since. This seems very possible, since these two tiny moons do not move round and round the equator of Jupiter; they cross its sky from north to south and south to north instead of from east to west. The outermost moon of Saturn and the one solitary moon of Neptune have similar motions, and it is just possible, although less likely, that either or both of these may have been captured in the same way. Even the smaller planets exert noticeable gravitational pulls, and the astronomer who tries to predict the future path of a planet or a comet must take all these pulls, small as well as big, into consideration.

The Discovery of the Outermost Planets

A century ago Uranus was thought to be the outermost planet of the sun's family. Astronomers had taken account of the gravitational pulls of the sun and all the other known planets, and had calculated the path it ought to follow. They found it did not keep strictly to its predicted orbit. They then began to suspect that some other, and hitherto unknown, planet must be pulling it out of its course. Two young mathematicians, an Englishman, J. C. Adams of Cambridge, and a Frenchman, U. J. J. Leverrier of Paris, set themselves the problem of discovering what and where the new planet must be, if its gravitational pull was to account for the erratic behaviour of Uranus. In due course the cause of the disturbance was found, almost exactly in the position which Adams and Leverrier had predicted. It is now known as the planet Neptune.

Recently history has repeated itself, and the same situation has recurred. Even after the gravitational pull of Neptune had been allowed for, Uranus still did not keep strictly to its predicted path, and astronomers began to suspect that yet another planet, out even beyond Neptune, must be pulling it from its course. This time it was an American, Professor Percival Lowell of Flagstaff Observatory, Arizona, who calculated how the supposed new planet must move. After fifteen years' search—and unhappily after Lowell's death—the planet was discovered in March, 1930, fairly near to where

Lowell had predicted it ought to be, and moving pretty much as he had predicted it ought to move. It is the newly discovered planet Pluto, which is about forty times as far from the sun as we are—so far out in space that its journey round the sun takes about 250 years to complete, and so far removed from the sun's light and heat that in all probability not only all its water but also its atmosphere, if it has one, must be frozen solid.

These two outermost planets, Neptune and Pluto, were both discovered as a result of the confidence astronomers feel in the law of gravitation, and their discovery provided a sufficient vindication for this confidence. If we are asked why we believe in the law of gravitation, perhaps the simplest answer we can give is that it enables us to discover new planets, although a more satisfying answer would be that it makes it possible to predict the motions of all the known planets.

Weighing the Stars

So far we have only dealt with the small colony of objects we call the solar system. Far out in depths of space—far beyond Neptune, Pluto, and the outermost confines of the solar system—we see other compact colonies. They are so far away that we should not see such small objects as planets and comets, even if they existed, but we see groups of stars which do not scatter but remain close neighbours in space. It is natural to conjecture that these, like the sun's family, are kept together by gravitation.

The colony which is believed to be nearest of all consists of three stars—two fairly bright, and one very faint (p. 97, below). But there are simpler colonies even than this. The simplest type of all, which we describe as a "binary system," consists of only two stars, each describing an orbit round the other—like two children holding both hands and dancing round and round, or like two partners in a waltz. They move exactly as they would if they were held together by the gravitational pulls they exert on one another, like the earth and moon, or the sun and earth. Thus we conclude it is gravitation that holds them together. The astronomer, watching the motion of the two stars round one another, can calculate how big a pull they must exert on one another to keep from separating, and in this way we learn the weights of some at least of the stars.

The results are interesting. Our sun proves to be of about average weight, or perhaps somewhat over. Taken as a whole, the stars shew only a small range in weight; if we compare the sun to a man of average weight, most of the weights of the stars lie between those of a boy and of a heavy man. Yet a few exceptional stars have quite exceptional weights. A colony of four stars, 27 Canis Majoris, is believed to have a total weight nearly 1000 times that of the sun, although this is not certain. An ordinary binary system, Plaskett's star, is believed, this time with fair certainty, to have a total weight of more than 140 suns. But such great weights are very exceptional. It is very rare to find a star with ten times the weight of the sun, and no star yet found has as little as a tenth of the sun's weight. Thus on the whole the stars shew only a very moderate range in weight.

The Candle-powers of the Stars

In contrast to their uniformity of weight, the stars shew a quite enormous range in candle-power. For instance, Sirius, the most brilliant star in the whole sky, has, close by its side, a quite dim star which only sends us about a ten-thousandth part as much light as Sirius does; it is so faint and so enshrouded in the glare of light from Sirius that it was not discovered until 1862. It is not a case of one star looking fainter than another because it is more distant, for this little star and Sirius form a binary system of the kind I have just described; the faint star does not move in a straight line through space, but goes round and round Sirius, shewing that it is permanently gripped by the gravitational pull of the brighter star. We can be sure, then, that the two stars must be at very nearly the same distance from us, and that the fainter star not only *looks* faint, but *is* faint—it is of low candle-power.

Even more sensational contrasts are known. The bright star Procyon has a faint companion which gives less than a hundred-thousandth part of the light of Procyon itself. Mira Ceti (or Omicron Ceti) also has a faint companion (see p. 92) which gives only about this same tiny fraction of the light of the main star. It is, then, certainly true that "one star differeth from another in glory," and this not merely because one star is further away than another.

Usually, however, we cannot compare the intrinsic brightnesses—the candle-powers—of two stars unless we know their distances. Only then can we tell how far apparent differences in brightness originate in mere differences of distance, and how far they represent true intrinsic differences of power.

As we know that the sun's distance from the earth is 92,900,000 miles, we can calculate what candle-power it must have to illuminate the earth as it does from this great distance. It must give out as much light as 3,000,000,000,000,000,000,000,000,000,000 candles.

Sirius is more than half a million times as distant as the sun. Light reaches us in eight minutes from the sun, but takes over eight years to reach us from Sirius. With this information before us, we can of course calculate the actual candle-power both of Sirius and of its faint companion. Sirius itself proves to be an unusually luminous star; it has about twenty-six times the candle-power of the sun, and its power as a radiator of heat is almost on a level with its light-radiating power. If Sirius were suddenly to replace the sun, our rivers and oceans, and even the ice continents round the poles, would rapidly boil away, and life would be banished from the earth. On the other hand, the faint companion of Sirius is of very feeble luminosity, even in comparison with the sun; it has only about a four-hundredth part of the sun's candle-power. If this faint star were put in place of the sun, and we had no other source of light and heat, the rivers and seas, even at the hottest parts of the earth's surface, would immediately freeze into solid ice, while our atmosphere would condense into liquid air.

Yet these two constituents of the system of Sirius are far from representing the utmost extremes we observe in the sky. The faintest star known, Wolf 359, is at least a hundred times feebler even than the faint companion of Sirius. At the other end of the scale is a "variable" star, S Doradus—a star, that is, whose light continually fluctuates in brightness. Its average candle-power is well over 10,000 times that of Sirius, and over 300,000 times that of our sun. When it is at its brightest, it has over 500,000 times the candle-power of the sun, so that it pours out as much radiation in a single minute as the sun does in a whole year. If the sun suddenly became as energetic as this star, its extreme heat would rapidly turn the whole earth and all objects on it, including ourselves, into vapour. If we compare the sun to a single candle, we must compare this star to an exceedingly powerful searchlight, while the faint star Wolf 359 must be compared to a very feeble glow-worm.

The Sizes of the Stars

The sun's radiation streams out uniformly from its whole vast surface, which is nearly 12,000 times as big as the surface of the earth. It is natural to wonder how large the surfaces are of the other stars, which pour out such different amounts of radiation. What, for instance, about S Doradus, when it is pouring out 500,000 times as much radiation as the sun? Has it got 500,000 times as big a surface as the sun, or does it pour out 500,000 times as much radiation from each square yard of surface, or what?

We may set about answering this question in one of two ways—either we may try to find out the size of S Doradus directly, or we may try to discover how much radiation it pours out from each square yard of its surface, and then deduce its size from the total amount of radiation it is known to emit. Unfortunately there are great difficulties in measuring the sizes of the stars directly. When we look at a planet in our telescopes, we see a round disc—like the moon, only smaller. If we could get near enough to the stars, they also would look like round discs, just as the sun does. But the sun is the only star which shews a disc of appreciable size, even in the most powerful telescopes. All the others are too far away for their discs to be observed; we only see them as mere pin-points of light, and so cannot measure their sizes directly.

There are, however, two exceptions to this general statement. The most ingenious instrument used by astronomers, the "interferometer," makes it possible to measure the actual sizes of a few of the largest stars by direct observation—it, so to speak, magnifies the tiny star discs in a very intricate way, until they become large enough to measure.

In another field, the most ingenious physical theory used by astronomers, Einstein's theory of relativity, makes it possible to measure the sizes of the smallest stars by direct measurement. So far this

PLATE XXVII

Mount Wilson Observatory

The Great Nebula in Orion

This is only the central portion of a vast nebula (see p. 120) which stretches over the greater part of the constellation of Orion. It will be recognised as an enlargement of that part of Plate XXVIII in which θ and the Great Nebula are indicated as lying.

PLATE XXVIII

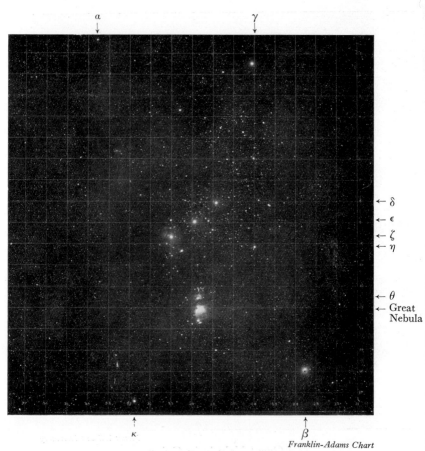

α γ

← δ
← ϵ
← ζ
← η

← θ
← Great Nebula

κ β

Franklin-Adams Chart

Part of the Constellation of Orion

The arrows indicate the principal stars. α is the huge red giant, Betelgeux (p. 84); β (Rigel) (p. 171) is one of the most luminous stars known, with about 15,000 times the candle-power of the sun; γ, and also δ, ϵ, ζ, the three stars of Orion's Belt, are intensely hot blue stars, as are also η and κ; θ, a quadruple star in the middle of the Great Nebula, is hotter still, with a surface-temperature of perhaps 50,000° Fahrenheit.
(Compare Plate III, facing p. 14.)

method has only been applied to one star, the faint companion to Sirius.

The number of stars whose size can be measured in either of these ways is, however, exceedingly small. Apart from these few, we have to approach the problem of how a star emits its energy from the other end, by trying to discover how much energy it emits from each square yard of surface. And here, happily, our way is clear.

The Colours of the Stars

To begin with, suppose that a football team is photographed, wearing its colours of red and blue. Everyone knows that the blue will come out nearly white in the photograph, while the red will come out nearly black. The reason is that the camera is very sensitive to blue and very insensitive to red, as compared with the retina of the human eye. Now we find that the camera plays similar tricks with the stars; photograph any bit of the sky we please, and some of the stars will come out unduly bright while others will come out unduly faint. The reason is of course that the stars are of different colours. Some are bluer than the average, others are redder; and the camera favours the blue stars and is quite unfair to the red. A conspicuous instance of this is shewn in Plate XXVIII, a photograph of part of the constellation of Orion. The faint star indicated by the arrow furthest to the left is Betelgeux, or Alpha Orionis. To our eyes, it appears the second brightest star in the constellation, and the twelfth brightest in the whole sky. Yet, because it shines with a deep red light, the camera exhibits it as a far feebler affair than the three stars of the belt. These latter look much fainter to our eyes, but as they happen to emit blue light, the camera sees them as very gorgeous stars indeed. This gives us an obvious method of discovering the colours of the stars, and it is found that the colour of a star can be told fairly accurately from the way it is treated by the camera. Other methods are available as well, and happily these confirm the story told by the camera.

The reason why stars are of different colours is that they are at different temperatures. As the blacksmith heats up a horse-shoe, its colour gradually changes—first it is dull red, then vivid red, then yellow and finally almost white, and the colour gives an indication of the temperature of the iron. In the same way, when a workman wishes to estimate the heat of a factory oven, his best rough-and-ready method is to look at the colour of the light it is sending out. A faint dull plum colour goes with one temperature, dull red with another, a bright red with another, and so on. Instruments are made to give the exact temperature of the interior of an oven from an examination of its light.

In precisely the same way, the astronomer can discover the temperatures of the stars. They exhibit a complete range of colours from a dull red, through yellow and white, to vivid blue and violet, and their range of temperature is correspondingly great. The dull red stars are the least hot, with temperatures of about 1400° Centigrade, or about 2550 degrees on the ordinary Fahrenheit scale. Yellowish stars are at least twice as hot. After these come stars like the sun, with temperatures of about 5500° Centigrade or 10,000° Fahrenheit, and so on until we come to the hottest stars of all, with temperatures of perhaps 70,000° Fahrenheit.

The whole observed range of temperature—from about 2550° to 70,000° Fahrenheit—is immense, and most of it lies entirely beyond anything we know on earth. Nevertheless, we can calculate how much radiation is given out by a given area at each temperature. The results are sensational. A surface at 70,000° Fahrenheit emits so much energy that power for all the railways of the world could be got from less than the standing-room for a single locomotive; each square inch of a surface at this temperature provides enough power to keep an ocean liner like the *Mauretania* going continuously at full speed. On the other hand, a square inch of surface at 2550° Fahrenheit, the temperature of the coolest stars known, hardly gives out enough power to propel a rowboat. Area for area, the hotter surface emits about 300,000 times as much radiation as the cooler. Consequently, a star at the lower temperature would need to have 300,000 times as big a surface as a star at the higher temperature, if it had to discharge an equal amount of radiation.

This of itself suggests that the stars must be of very different sizes. If the dull red stars are to be of reasonable candle-power, they simply have to be enormous; their candle-power per square inch is so small. In actual fact some of these dull red stars are

6-2

of quite terrific candle-power, and give out great amounts of heat as well. For instance the total radiation, light and heat added together, of the star Betelgeux, or Alpha Orionis, which we mentioned just now, is about 6000 times that of the sun. As the star is dull red in colour, it can emit nothing like as much radiation per square inch as the sun, so that its surface must be far more than 6000 times as big as that of the sun.

If we can find the exact temperature of Betelgeux from its colour, we can calculate how much radiation pours out from each square inch of its surface, and so can discover how many square inches of surface it needs to emit its observed total radiation—in brief we can tell how big it is. The same method is of course applicable to any other star. First, its observed temperature tells us its output of radiation per square inch. And then, if we know its total radiation, a simple division gives us the total number of square inches of its surface.

We have already noticed the two ways in which the sizes of a few stars can be measured directly. Whenever the size of a star has been measured in either of these ways, the result has come out very near to the calculated value—the value we obtain by dividing the total radiation of the star by its radiation per square inch. This gives us every right to feel confidence in our method of calculation.

Now such calculations give very sensational results. They shew that the stars vary far more in size than in weight or temperature, and more even than in candle-power. The smallest star so far discovered, van Maanen's star, is hardly, if at all, greater than our earth; a million such stars could be packed inside the sun and leave room to spare. This makes the sun seem quite a big star, yet there are other stars, such as Betelgeux, so enormous that many millions of stars of the size of the sun could be packed inside them and leave room to spare; they are so big that if one of them were put in the position of the sun, we should find ourselves inside it, the radius of the star being greater than that of the earth's orbit. Let us again imagine that the sun is represented by a pea, then the tiniest of stars, such as van Maanen's star, will be represented by specks of dust so small that eighty of them would barely cover the dot of an "i" in this type, while the largest stars must be represented by globes as big as motor-cars.

We see that the museum of the sky contains a vast range of exhibits, and we cannot but wonder what is the origin and meaning of so amazing a variety. Why are the stars so much alike in their weights, and so little alike in everything else? To this question we shall turn in our next chapter.

CHAPTER V

THE VARIETY OF THE STARS

We have seen how the stars shew as great a range oᵢ candle-power as there is between a glow-worm and a searchlight; while their range of size is as that between a speck of dust and a motor-car. The range in their weights is much smaller, but still it is about equal to that between a feather and a football. And in every respect the sun is somewhere about average. It could hardly be expected to strike the exact happy mean in every way, but it never misses it badly. To put the same thing in another and less complimentary way, the sun is totally undistinguished in all respects—in weight, in size, in temperature and in candle-power.

Clearly, however, we get very little knowledge of the general nature of the stars from a mere mention of extremes and of one average star. We should not know much about the English population if we had only been told the heights and weights of the shortest dwarf and the tallest man, and that a particular man 5 feet 9 inches high was a good average Englishman in all respects. We want a more detailed knowledge as to the classification of the stars by size, candle-power and weight.

Suppose that all the entrants to a dog-show broke loose and ate their labels, and had to be reclassified. A novice would probably think it necessary to classify them several times over, first by weight, then by colour of coat, then by length of coat, and so on; the expert would at once set to work to classify them by breeds. The weight, colour and coat of individual dogs would vary substantially inside each breed, but nothing like so much as through the whole range of dogs.

Three Types of Stars

It is much the same with the stars. They look an exceedingly mixed lot to the casual observer, but the expert astronomer knows that they can be classified into distinct types with almost the same precision as dogs at a dog-show. Indeed there are almost innumerable breeds of dogs, but only three main types of stars, which are distinguished primarily by their sizes. We must not compare the stars to a whole dog-show broken loose, but only to three classes, one of very small, one of medium-sized, and one of immense, dogs. Of course the comparison is not perfect; the sky is not quite so simple as this. The main respect in which it fails is that there is a gradual transition between the two largest classes, although not, so far as we know at present, between either of these and the class of very small dogs.

Before trying our hand at classifying actual stars into these three types, let us try to understand how they come to exist. First of all, why are there distinct types of stars at all? Our knowledge of the structure of the atom seems to provide at least a partial answer.

During our rocket-tour inside the sun, we noticed that an ordinary atom consists of a central nucleus with a number of tiny and almost weightless electrons arranged round it. At such temperatures as we experience on earth, the nucleus holds all its electrons in a firm grip. With greater heat, such as we found in the sun's atmosphere, the outermost electrons begin to work loose. Finally, at the extreme heat of the sun's centre, we found that all the electrons had worked loose except for an inner ring of two which are held in a very specially firm grip, a grip so tight that it can defy a temperature even of 40 million degrees.

White Dwarfs

Yet stars are known whose centres are ten, twenty—perhaps even fifty—times hotter than the centre of the sun. No nucleus can grip its electrons strongly enough to defy such heat as this. At the centres of these stars every atom is completely broken up; the only substance there consists of what we may call powdered atoms, a mere disordered crowd of nuclei and electrons dashing about helter-skelter in all directions without any attempt at coherence—matter in its most rudimentary form. As we have no experience of such a state on earth, it is difficult to find a single word with which to describe it. It is like a gas in consisting of a number of tiny particles each of which moves independently of the rest. Yet the particles are jammed so closely together that perhaps we shall get a better picture if we compare the substance to a liquid such as water or quicksilver.

A complete unbroken atom is like the solar system in miniature; the massive central nucleus is the sun, and the electrons are the planets. And again, it is like the solar system in consisting mainly of empty space. We have already seen how tiny the sun and planets are in comparison with their distances apart. We made a model of their arrangement by laying out a pea, two small seeds, some grains of sand, and some specks of dust in Piccadilly Circus. The whole of Piccadilly Circus was needed to represent the *space* of the solar system, but a child can carry the whole *substance* of the model in its hand. All the rest is empty space. It is the same with the atom. If we take Piccadilly Circus to represent the space occupied by the atom, its material ingredients —the nucleus and the electrons—will at most be a few small seeds, which again can be packed in a very tiny space.

At the centres of the hottest stars of all, the tiny ingredients of the atoms are so packed. After the fierce heat has broken the atom into its constituent nucleus and electrons, the intense pressure, produced by the weight of all the rest of the star, comes into play and jams these ingredients close together. This packs the star's substance into an amazingly small space, so that the star is of very small size.

This way of packing a star's substance gives us the smallest class of stars, the class which astronomers describe as "white dwarfs." An extreme instance is van Maanen's star (p. 85), which is no larger than the earth. A less extreme example is the faint companion of Sirius (p. 77). This is about thirty times as big as the earth, but as it contains 300,000 times as much substance, it must be packed 10,000 times as closely as the earth. We see that Nature can still teach us something in the art of packing. If we could pack our terrestrial goods as closely as these stars are packed at their centres, we could carry about 100 tons of tobacco in a tobacco-pouch, and several tons of coal in each waistcoat pocket. Compared with the powdered atoms of which these stars are formed, the solid substance of our earth is but the finest of gossamer—a sort of spider-web spun in space.

Because the stars of this type are so compact, each bit of their minute surfaces is called on to radiate a vast amount of energy. Generally speaking, every square inch of surface gives out about 250 horse-power of energy, as against 50 horse-power per square inch for the sun. To get rid of all this energy, the star's surface must be at a white heat. We now see why stars of this type are known as "white dwarfs" —dwarfs because of their small size, white because they are white-hot

Main-sequence Stars

The white-dwarf stars are rather exceptional, and the substance of most stars is not so closely packed. When we were inside the sun, we noticed that the majority of the atoms were not completely broken up; many of their nuclei still retained one or two electrons, thus forming real atoms of a definite, although very minute, size. Such atoms cannot be crammed into almost no space at all, like the substance at the centres of the white dwarfs, but they can be packed into very much less space than unbroken atoms could be. They are packed in this way at the centre of the sun, so that a cubic foot of substance contains some tons—we do not quite know how many. At the centre of a white dwarf, a cubic foot contains many thousands of tons.

The sun, with its substance packed in this way, is representative of the largest class of stars, the stars of medium size, which are known as "main-sequence" stars. This class comprises perhaps 80 per cent. of all the stars in the sky. The centres of all main-sequence stars are about as hot as the centre of the sun, with the result that the atoms usually retain their two innermost electrons and no others—we may compare them to solar systems in which only Mercury and Venus are left describing their orbits. And although matter which is broken up to this extent can be packed very closely, it cannot be packed anything like as closely as it is packed in the white dwarfs. The result is that the main-sequence stars are all substantially larger than any of the white dwarfs, and do not shew any enormous range of sizes. But apart from their sizes they shew the utmost diversity, their weights ranging through the whole gamut of known stellar weights, and their colours through the whole spectrum of known colours from the most vivid violet to the dullest red. Yet, as the name "main sequence" implies, they form a true sequence. When we arrange them in order of weight, we find that we have also arranged them very approximately in order of colour; their heaviest stars are also their bluest, and with decreasing weight, the colour passes steadily down the spectrum, through blue, white and yellow, to the dullest and deepest of reds. And again, with decreasing weight the candle-power continually decreases, running through the whole range of stellar luminosities from the searchlight to the glow-worm.

Red Giants

Stars of the third class are distinguished by their centres being much cooler than the centres even of main-sequence stars; the temperature may be as low as one or two million degrees. In such comparative coolness as this, the electrons are not stripped off the atom down to the innermost ring of two, as they are in the sun. Other rings of electrons are left clinging to the nucleus, so that the atoms remain of very respectable size and cannot be packed very close together. Indeed, they seem to make themselves very comfortable, retaining so much elbow-room that the stars we are now discussing are of immense size. A typical example is the star Betelgeux, or Alpha Orionis, which is 25 million times as big as the sun, although probably containing only about forty times as much substance. A still larger example is Omicron Ceti, which is so large that 30 million suns could be packed inside it. This has been recently found to have a faint companion in the shape of a white dwarf, the two together forming a binary system. If the stars had a sense of humour, this ill-assorted pair ought to create some merriment by their absurd incongruity and inequality of size. Landseer's "Dignity and Impudence" is quite outdone; it is as though an elephant and a sandfly were to join hands and travel about through space in one another's company.

Most of these stars are big enough to hold at least a million suns in their interiors. Although their candle-power is terrific, yet they have such immense surfaces that each square inch has only a small amount of energy to dispose of; sometimes as little as half a horse-power, as compared with 50 horse-power in the sun, and 50,000 horse-power in some blue main-sequence stars. The surface can dispose of this small amount of energy without becoming unduly hot, so that its colour is generally red or, less often, yellow.

These stars may be described as red and yellow giants—giants because of their size, and red or yellow because of their colour.

Stellar Energy

It seems pretty clear that the wide difference in the size of the stars is associated with a corresponding difference in the size of the atoms in the star's interior, but we understand the wide range in candle-power less well. It seems fairly certain that each star must be regarded as a huge power-station, which generates energy in its interior, and pours it out into space in the form of radiation from its hot surface. The sun's output of 50 horse-power per square inch may seem tremendous at first glance; but we must remember that a square inch of surface is the only outlet for the energy generated in a large bulk of star. As the sun has a radius of 432,000 miles, all the energy generated in 432,000 miles of substance behind a square inch of the sun's surface has to pour out through that square inch. When we look at it this way, 50 horse-power per square inch does not seem excessive, rather the reverse.

We know that radiation has weight, so that a stream of weight must continually be pouring out of every bit of a star's surface. Calculation shews that the total radiation which pours out from the sun every second has a weight of 4 million tons. Thus the sun must continually be losing weight at the rate of 4 million tons a second, which is about 10,000 times the rate at which water flows under Westminster Bridge. Its weight is diminishing as surely as if there were 10,000 gashes in its surface, with a whole River Thames pouring out of each. At this moment it weighs many millions of tons less than when you began reading this chapter; by this time to-morrow it will weigh 350,000 million tons less than now. Where does all this weight come from?

The Stars destroying their Substance

We still do not know for certain how a star generates its radiation, but it seems most likely that it does it by destroying its own substance, just as an ordinary power-station generates energy by burning coal. But the process going on in the star is something very different from a mere combustion, which only involves a re-arrangement of atoms. More probably it is an actual annihilation of atoms: an atom exists at one instant, and the next it is snuffed out: nothing remains but a flash of radiation, which, however, will have precisely the weight of the atom which has disappeared.

If this is the true origin of the radiation of the stars, the sun is destroying its atoms at the rate of 4 million tons a second, or 350,000 million tons a day, and other stars must be destroying their atoms at other, but comparable, rates proportional to their various candle-powers. Because of the weight they continually lose in the form of radiation, the stars must be continually getting lighter in weight, so that, broadly speaking, the oldest stars ought to be those of lightest weight. There is a good deal of evidence to suggest that such is actually the case.

We have already seen how the stars of greatest weight, which we must now regard as the youngest, are far and away the most luminous, and generally speaking, the candle-powers of the stars fall off with diminishing weight. But they lose in candle-power far more rapidly than in weight. The old star not only has less substance left, but what is left is less efficient in radiating power per ton. We can best explain this by supposing that a star consists of a mixture of substances which turn themselves into radiation at different rates. Some substances transform themselves very quickly, and so give off radiation very rapidly while they last, but do not last for long. While they last the star radiates furiously. After they have been used up, the feebler kinds which are left give off radiation more slowly, and so last for much longer, so that, after a brief but boisterous youth of squandering its substance at a profligate rate, a star can look forward to a serene and protracted old age in which it radiates its energy more sedately. Although this scheme cannot be said to be finally and unalterably established, it accords well enough with the known facts of astronomy, and it will at least serve to give interest to the great variety of stars observed in the sky.

The nearest Stars

With the foregoing classification of the stars in our minds, let us briefly survey our nearest neighbours in space. These probably give us a fairly good sample of the sky. If we went further out into space, we should quite certainly get a worse sample, because we should then be ignoring great numbers of very faint stars, which are unknown and unobserved as the result of their combination of distance with faintness. It is only quite near home that we can count on very faint stars being observed. A list of our 26 nearest neighbours in space, with their distances in light-years, is given opposite. The circles which follow in the next column give (approximately) the relative sizes of the stars, and this is followed by their colour as seen through the earth's atmosphere. Finally the last column shews the approximate radiating powers of the stars, that of the sun being taken as unity.

Stars	Distance in light-years	Size and colour		Candle-power (in terms of sun)
Sun	—		Yellow	1
Prox. Centauri	4·27		Red	$\frac{1}{20000}$
α Centauri	4·31		Both yellow	$1\frac{1}{8}$ and $\frac{1}{3}$
Munich 15040	6·06		Red	$\frac{1}{2500}$
Wolf 359	8·07		Red	$\frac{1}{50000}$
Lalande 21185	8·33		Red	$\frac{1}{200}$
Sirius	8·65		Both white	26 and $\frac{1}{400}$
3 very faint stars	9 to 10		?, ?, Red	Average $\frac{1}{900}$
τ Ceti	10·3		Reddish yellow	$\frac{1}{3}$
Procyon	10·4		White, ?	$5\frac{1}{3}$ and $\frac{1}{20000}$
8 faint stars	$10\frac{1}{2}$ to $11\frac{1}{2}$		All red	Average $\frac{1}{10}$
Kruger 60	12·7		Both red	$\frac{1}{400}$ and $\frac{1}{1400}$
Van Maanen's star	12·8		White	$\frac{1}{6000}$

Assuming that these 26 stars are typical of the whole sky, we notice at once that the majority of stars are both redder and smaller than the sun, and so of course must also be less luminous. Probably only four of the 26 stars are larger than the sun, while only three, the brighter components of the systems of Alpha Centauri, Sirius and Procyon, are more luminous.

We notice that the whole group of stars does not contain a single red or yellow giant. This does not mean that the vicinity of the sun is in any way abnormal. Giant stars are extremely rare in space, so that there are immense odds against even a single one occurring in any small collection of stars. If there had chanced to be a red or yellow giant in the vicinity of the sun, we should not have been able to exhibit it in our diagram: an average red giant would be represented by a circle 12 feet in diameter. Out of these 26 stars, 23 are certainly main-sequence stars, while one, the faint companion to Procyon, is doubtful—it may be a white dwarf. The remaining two, the faint companion to Sirius and van Maanen's star, are certainly white dwarfs. Our sample is adequate to shew that the vast majority of stars in the sky belong to the main sequence.

These 26 stars are turning their substance into radiation at different rates, but most of them more slowly than the sun; only three, one in each of the systems of Alpha Centauri, Sirius and Procyon, are wasting away more rapidly than the sun, and these all have more substance to waste. The sun's present store of atoms would last it for about 15 million

million years at its present rate of consumption, but long before it comes to its last atom, it will have reached the state of the smaller and fainter stars, and be radiating its substance away far more slowly than now.

Allowing for considerations of this kind, it seems probable that most stars may look forward to lives of hundreds of millions of millions of years before darkness finally overtakes them. Whether this estimate becomes finally established or not, one thing seems certain—our human lives fade into utter insignificance when measured against astronomical time. We have seen how the earth is only a speck in space; we now see that our lives, and indeed the whole of human history, are only a speck in time.

CHAPTER VI

THE MILKY WAY

When we discussed the face of the sky in our first chapter, the stars were nothing to us but a distant background of points of light. This background enabled us to fix our bearings in space, and we saw how we could pick out our near neighbours, the planets and other members of the sun's family, by their rapid motion against it.

Since then we have examined what the stars really are, and have discussed their various physical characteristics. Amongst other things, we have found that they shew great variety in their candle-powers. While some are thousands of times more luminous than our sun, others are thousands of times fainter. If we compare our sun to an ordinary candle, some stars must be compared to searchlights, and some, at the other end of the scale, to glow-worms or fireflies.

It has only recently been discovered how great a range there is in the candle-powers of the stars. For a long time it was supposed that the stars all had pretty much the same intrinsic brightness—like a row of street lamps—so that when a star looked very dim, it was only because it was very distant. In 1761 the astronomer Lambert argued that as all the stars had been made to serve the same purpose, there was no reason why some should have been made fainter than others; if some appeared fainter it could only be that they were more distant. We have already seen how utterly and completely wrong this conclusion was.

Mapping out the Universe

If Lambert had been right, and all the stars had been found to have the same intrinsic brightness, like a row of street lamps, astronomy would have been a much simpler science than it is. We could at once have deduced a star's distance from its apparent brightness, and so could have mapped out the universe star by star. As things actually are, the faint star we are looking at may be a very distant searchlight, or it may be a very near glow-worm. It is hard to say which; we can only decide by measuring its distance.

We have seen how we can measure the distances of some stars by the ordinary surveyors' method, noticing how much they change their positions as we move about in space. But this only applies to a few very near stars. The longest journey Nature permits us to take is the journey of 186 million miles which we take every six months as the earth passes from one side of the sun to the other. And most of the stars are so remote that even this long journey causes no appreciable change in the directions in which we see them. We are in effect faced with the problem of measuring the distances of objects by merely looking at them, without being allowed to move about. How are we to do it?

We have seen how we could do it with a row of street lamps, provided we knew they were all of the same candle-power. We use the same method for the stars. As a whole, they are of widely different candle-powers, but it has recently been discovered that certain classes of stars, easily recognisable in themselves, are of a uniform standard candle-power. As soon as we know the candle-power of one, we know the candle-power of them all, and then we can use the "street-lamp" method to judge their distances—the fainter the star looks, the more distant it is. Or, even more concisely, the star is just as distant as it appears to be.

The street-lamp method would of course fail if there were any sort of fog or obscuring matter pervading space, and extinguishing light after it had travelled a certain distance. On a foggy night, we can only see a few of the very nearest lights in a street, and must not judge their distances from their apparent faintness; the faintest are not as distant as we might think them, if we did not know that we were looking at them through a fog. Very careful investigations seem to shew that there is no such fog in space, except in a few special directions. Scattered here and there over the sky are a number of clearly defined black patches in which we see either no stars at all, or else a few whose brightness proves them to be quite near. A conspicuous instance is the jet-black patch known as the Coal Sack, which appears near the middle of Plate XXIX. These patches look like yawning cavities and used to be interpreted as such; they were thought to be holes in the system of stars—a system of tunnels reaching from outer space to the earth, although it must always have seemed

PLATE XXIX

W. H. Steavenson, Pretoria

The Milky Way—I

This plate shews the southernmost part from Centaurus (top) to Argo (bottom). The bright stars near top (centre) are α and β Centauri. Below them is the Coal Sack (pp. 102, 178). To the right of this is the Southern Cross, and lower down is the nebula surrounding η Argus. A quarter down near the right-hand edge is the globular cluster ω Centauri.

PLATE XXX

Mount Wilson Observatory

Nebulosity in Orion

The "horse's-head" south of ζ Orionis—"the smoke of our own star-city, lighted up by the lights of our own star-city" (see p. 119).

very odd that so many tunnels should converge on our tiny earth. We now know that the black empty spaces are not tunnels at all—they are clouds of dark matter, fairly near home, which prevent us seeing the stars beyond.

A mere inspection of modern photographs is enough to establish this interpretation. For instance, the dark patch shaped like a horse's head near the middle of Plate XXX, cannot possibly be interpreted as a tunnel through the stars; we see at once that it is some kind of obstruction.

Except in the few directions in which we meet with obscuring matter of this kind, astronomical space appears to be perfectly transparent. Starlight travels through it without let or hindrance, and suffers no diminution except on account of distance. So that of any special classes of stars which are of uniform candle-power, it is strictly true that a star is just as distant as it appears to be. The most interesting stars of this kind are a class known as Cepheid variables.

Cepheid Variables

The light of most stars is perfectly steady, but there are a few rare stars whose light continually fluctuates from bright to faint and back again, as though someone were continually turning the tap of a gas-lamp up and down. Long ago a star, Delta Cephei, was observed to fluctuate in a very peculiar way—as though the tap were turned down gradually, and then suddenly turned up again with a rush. It repeats this cycle of changes with the utmost regularity every

$5\frac{1}{3}$ days. A distant star-cloud, the Lesser Magellanic Cloud (see Plate XXXI), contains a whole lot of stars exactly similar to this, and it is found that they all look equally bright. As they are all at the same distance, this means that they must all have the same candle-power. Other stars of precisely the same kind are found so near home that we can measure their distances by the ordinary surveyors' method, and so can of course calculate their actual candle-powers. All these, too, are found to have the same candle-power. To sum up a whole lot of astronomical research, it is found that every star which behaves like Delta Cephei has the same candle-power as Delta Cephei.

Other stars shew characteristic light-fluctuations of the same general kind—the same gradual dimming followed by a rapid restoration of brightness—but have periods of fluctuation different from the $5\frac{1}{3}$ days of Delta Cephei. All such stars, as a class, are known as Cepheid variables. Again all those which have the same time of fluctuation, whatever it is, are found to have the same candle-power, which we discover, as before, by calculating the candle-power of a star near home. Thus we can tell the candle-power of any Cepheid variable in the sky by noticing the time of its fluctuation, and then we can deduce its distance from its apparent brightness. These stars are like lighthouses in the vast oceans of space. We recognise them at once and unmistakably by the characteristic fluctuations of their lights, and knowing their candle-powers, we can at once deduce their distances.

PLATE XXXI

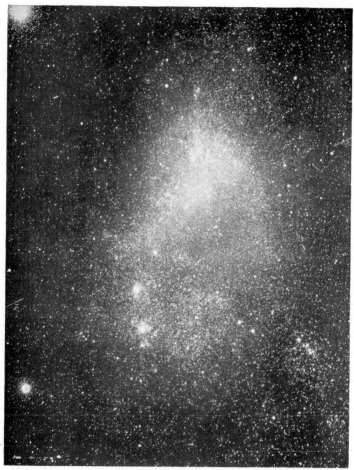

Arequipa Observatory

The Lesser Magellanic Cloud

This vast cloud of stars lies in the constellation Tucana, near the South Pole, and so cannot be seen in England. It is so large that light travelling at 186,000 miles a second takes 6000 years to pass from one end of it to the other, and so distant that its light takes 95,000 years to reach us. It contains at least 500,000 stars brighter than Sirius, as well as immense numbers of fainter stars, and yet, owing to its great distance, we receive only a twenty-fifth as much light from it as from Sirius.

Two globular clusters can be seen near the left-hand edge of the plate. That at the extreme top is 47 Tucanae, one of the nearest and brightest of all the globular clusters at only a fifth of the distance of the Magellanic Cloud (see p. 106).

PLATE XXXII

Dominion Astrophysical Observatory, Victoria, B.C.

The Globular Cluster M 13 in Hercules

This is the finest cluster visible in the Northern Hemisphere, although not in the whole sky. Its distance is such that its light takes 33,000 years to reach us. Although it sends out 2½ million times as much light as the sun, yet it is barely visible to the unaided eye.

This gives us a most valuable method for sounding space, or at least those parts of space in which we can see Cepheid variables. Shortly after the method was discovered, the present Director of Harvard Observatory, Dr Shapley, used it to measure the distances of certain clusters of stars, known as "Globular Clusters," each of which contains some hundreds of thousands of stars.

Globular Clusters

Imagine a swarm of bees settling in the open air. They form a central globular mass, round the outskirts of which an enormous number of bees buzz about and form a sort of atmosphere to the main swarm. If we replace each bee by a star, we shall have a very fair representation of what a globular cluster looks like—a round mass of stars, whose members are closest together at the centre and more widely spaced on the outskirts. A typical cluster is shewn in Plate XXXII.

About a hundred of these globular clusters are known. No new ones are being discovered now, or have been discovered in the last century or so, so that we may take it that there are none left to discover; we know them all. They mostly look very faint objects in the sky, only five or six being visible to the unaided eye.

They all contain great numbers of Cepheid variables, and this makes it possible to estimate their distances with considerable accuracy, and with sensational results. Even the nearest of the globular clusters proves to be so remote that its light takes about 18,400 years to reach us. We do not see it as it now is, or where it now is; we see it where it was, and as it was, 18,400 years ago—long before man became civilised. We see it by light which started out on its long journey to us while the earth was still covered with primaeval forest and overrun with wild beasts, while agriculture was unknown and man lived by the crudest kinds of hunting and fishing. While this light has been travelling through space on its way to us, all the recorded history of mankind has taken place; 600 generations of man have been born, lived their lives and died; empires have risen, decayed and fallen—it has taken all this time for the light from even this nearest of the globular clusters to reach us, travelling through space at a rate of more than 11 million miles every minute. This cluster contains hundreds of thousands of stars, many of which are far more luminous than the sun. And yet it is so distant that it can only be faintly seen by the unaided eye.

If this cluster should contain astronomers amongst its inhabitants, who study us as we study them, they will see the earth's yearly path round the sun looking the size of a pinhead 400 miles away. This shews how useless the old surveyors' method of measuring the distances of the stars becomes at these great distances. Beings who can only crawl round a pinhead cannot expect to see objects 400 miles away changing their positions by a measurable amount.

Shapley further found that the most distant cluster

is about ten times as remote as the nearest; while the light from the nearest cluster takes about 18,400 years to reach us, that from the furthest takes about 185,000 years. He also measured the distances of all the intermediate clusters, and mapped out their positions in space. Their general arrangement in space proved to be rather like the arrangement of currants in an ordinary currant-bun; in other words they are fairly uniformly distributed through a space shaped like a bun, a space of circular cross-section whose thickness is less than its length and breadth.

Although this is not yet certain, it now seems likely that in mapping out the globular clusters in this way, Shapley was solving a bigger problem than was realised at the time; he was solving the problem of the arrangement of the stars in space.

Probably the first instinct of primitive man was to suppose that the stars went on for ever and ever. It is the simplest, and in many ways the most natural supposition to make. Yet a great many considerations shew that it cannot be the correct one. To mention only one, if the stars went on for ever and ever, arranged as they are near the sun, then, in whatever direction we looked, we should be certain to come upon a star sooner or later. The whole sky would appear as a uniform unbroken blaze of light, just as in a blizzard the whole sky appears as a uniform unbroken sheet of snow. As the greater part of the night sky is black, we may be sure that the stars do not go on for ever; after we have gone a certain distance into space they must begin to thin out and finally disappear. Apart from the special regions already mentioned, in which star-light is blocked out by patches of dark matter, the sky only appears black where we have looked right through the whole system of stars into the vacant space beyond.

The Milky Way

Yet not quite all of the night sky is black. On any clear moonless night, we see a great arch of faint pearly light, spanning the sky from horizon to horizon. We cannot see what happens to it below the horizon, unless we travel round the world. We then find that its two ends join up in the southern sky, so that it forms a great endless circle of light which goes round the whole sky—a belt of light encircling the world. In nearly all languages it has the same name —the Milky Way.

Not only primitive peoples but even the early astronomers were much mystified as to the nature of this arch of light. The Mexicans called it very poetically "The little white sister of the many-coloured rainbow," and with most civilisations it formed the subject of innumerable legends—you may remember Tintoretto's picture "The Origin of the Milky Way" in the National Gallery (see the Frontispiece). Then in 1609 Galileo turned his newly constructed telescope upon it, and the mystery was immediately solved. The Milky Way was seen to be nothing but a cloud of faint stars, scattered like a fine silvery dust on the velvety background of the sky (see Plates XXIX, p. 102, and XXXIII, p. 114). Galileo's telescope further shewed that, even in the

Milky Way, the greater part of the sky is blackness, stars being only incidents on a black background.

Except where there are patches of obscuring matter, this blackness can only result from our looking through the whole system of stars into the vacant space beyond, so that even in the direction of the Milky Way, we come to the end of the stars in time. Yet many more stars can be seen in this direction than in any other; also they look fainter, which suggests that they are more distant. Clearly we can travel much further in this direction than in any other before coming to the end of the stars.

The Cart-wheel of the Stars

Sir William Herschel reached this conclusion 120 years ago. He thought that the stars were arranged rather like an enormous cart-wheel, with the sun somewhere near the hub. He supposed that the stars in the rim of the wheel formed the Milky Way; the stars in this direction looked faint because they were so distant, and they appeared specially numerous because, when we looked in this direction, we saw not only the stars of the rim but the stars all along the length of a spoke as well.

Recent work in astronomy has confirmed Sir William Herschel's conclusions in many respects, but shews that in one respect he was wrong: the sun is not at, or even near, the hub of the great wheel of stars, but part way along a spoke—perhaps something like a third of the way from hub to rim. For we now know that this great wheel of stars is rotating in space. It is not rotating round the sun, or round a point anywhere near the sun, but round a hub so distant that light from it takes about 50,000 years to reach us. This hub lies in almost exactly the same direction as the centre of the bun in which we have imagined the system of the globular clusters to be enclosed, and is at about the same distance. Also the plane of the wheel, which is of course the plane in which the Milky Way lies, agrees exactly with the central plane of the bun; half of the globular clusters lie on one side of the Milky Way, and half on the other.

This establishes, almost beyond doubt, that Sir William Herschel's round cart-wheel is essentially the same thing as the round bun by which we have represented the arrangement of the globular clusters in space; the stars occupy the same region of space as the globular clusters, and as we travel out into space the two leave off more or less together. There is only one difference: the cart-wheel which represents the stars is not quite so thick as the bun which represents the globular clusters. Perhaps we can put it better as follows.

Let us butter our bun. Let us cut it into two halves, top and bottom, spread a good thick layer of butter between the two halves, and then put them back into position again. Then the butter represents the stars, and the currants the globular clusters. The sun is not, as Sir William Herschel thought, near the middle of the bun; it is true that there is as much of the bun above it as below it, so that it comes in the middle of the layer of butter, but it is nearly halfway out from the centre to the rim.

This very prosaic model is the simplest I can devise to explain the arrangement underlying the majestic magnificence of the night sky. To pass from the model to the reality, we must magnify and magnify and magnify, until each tiny speck of space becomes millions of miles: we must replace each currant by a cluster of hundreds of thousands of stars, our layer of butter by a cloud of many millions of stars, and let all else dissolve into the velvety blackness of empty space, or, at most, into sparsely scattered atoms, or broken fragments of atoms, and clouds of dust. If we can persuade our imaginations to make all these transformations for us, the result will be anything but prosaic; it will provide us with a key to the most impressive spectacle which the eye of man ever beholds. It will enable us to look at the wonderful panorama of the sky with a new understanding of what it all means.

The Night Sky

Yet, even so, we must not expect to see the whole structure spread before our eyes when we stand in the open and look up to a clear night sky. The distances in space are so immense that even the most luminous stars only affect our unaided eyes if they happen to be comparatively near to us. Unless their light can reach us in less than about three thousand years, we shall not see them without optical aid. Now even the nearest of the globular clusters is six times as far away as this. Thus we may say that all the stars we can see individually, *as stars,* lie in a very tiny bit of space surrounding our sun—in a bit of our currant-bun which is itself not much larger than a good-sized currant. If every star outside this tiny bit of space were suddenly annihilated, our unaided eyes would not see a single star disappear from view. The Milky Way would disappear, because it is formed by the combined light of a great multitude of stars which are too far away from us to be seen individually—like the lights of a distant town. And the general background of the sky would become just a bit darker, because a faint and almost imperceptible haze of light is spread all over this by distant stars, which again are too remote to be seen individually. But our unaided eyes would discern no other changes. All the stars we see separately as stars would remain unaffected; they are all very near home indeed, when we measure distance on the astronomical scale.

And so it results that our view of the night sky falls into two distinct parts. First, we see the constellations, which consist of a very near foreground of separate stars—that is to say, near on the astronomical scale. Second, we see the Milky Way, which is a background formed of stars so distant that we only see them as a crowd. Constellations and Milky Way—these are all that we see. In the middle distance between are millions of stars which we do not see at all, because they are too distant to be seen as separate stars and too few to appear as a continuous cloud of light; at best they merely lighten up, in small measure, the dark background of the sky.

This whole system of stars, the wheel-shaped

system bounded by the Milky Way as rim, is usually described as the "Galactic System."

The Number of the Stars

If we could see all the stars of the Galactic System as separate stars, how many would there be? At first this may seem the simplest of all the questions the astronomer has to answer; surely he has only to count them in his telescope. Unfortunately it is nothing like so simple as that. The larger the telescope the more stars we see. The largest telescope yet built shews about 1500 million stars—roughly a star for each inhabitant of the earth who is more than five years old. But a still larger telescope which is now under construction will almost certainly shew many more, and even then we cannot expect to see all the stars, or anything like all. No; it is no use trying to count the stars—there is only one way of telling their total number, and that is to weigh them all together.

It may sound crazy to talk of weighing stars we cannot even see, but this describes quite literally what astronomers have recently been doing.

For a long time there was some doubt as to how the system of stars was able to retain its shape of a disc or a wheel. It was hard to see why the gravitational pull of the stars at the hub did not draw in the stars of the rim until they all fell together in a bunch at the centre. This puzzle is now solved; the wheel retains its shape simply because it is revolving round the hub. In this it is rather like the solar system, but on a stupendous scale. The solar system is also shaped like a disc or a wheel. There is no mystery as to why it retains this shape; it is because the planets are revolving round the sun. They would fall in if they stopped revolving; actually their motion round the sun saves them from this fate. The planets which are nearest the sun have to move fastest, because the sun's gravitational pull against which they have to contend is strongest where they are. It is the same in the far larger system of stars: their motion saves them from falling into the hub. The gravitational pull is strongest near the hub, so that the stars which are nearest the hub move fastest. The sun, at some distance from the hub, moves at something like 200 miles a second, which is 10,000 times the speed of an express train. And its distance from the hub is so great that, even moving at this speed, its journey round the hub probably takes two or three hundred millions of years.

These figures are not at all exact; we still do not know with any accuracy how far we are from the hub round which we are turning. We have a far better knowledge of the direction in which this hub lies. It must of course lie in the Milky Way, and is almost certainly in the stretch shewn in Plate XXXIII. Probably it is somewhere near the middle of it.

Now the middle of this stretch has long been known as the richest part of the Milky Way. We should expect the stars to cluster most thickly round the hub of the wheel, and in any case we should see the greatest depth of stars by looking in the direction of the hub to the rim beyond, so that it is not surprising that the hub should be found to lie in a rich part of the Milky Way.

PLATE XXXIII

W. H. Steavenson, Pretoria

The Milky Way—II

This plate shews the brightest and widest part of the Milky Way, from Antinous to Scorpio. The middle part is shewn in more detail in Plate XXXIV (overleaf).

PLATE XXXIV

E. E. Barnard, Yerkes Observatory

The Milky Way in Sagittarius

This is the central part of Plate XXXIII in greater detail. The left and centre of the plate shew the Great Star-Cloud in Sagittarius, the brightest part of the Milky Way. The dark obscuring patch to the right (seen more fully in Plate XXXIII) probably conceals the hub of the Galactic System.

The richest part of all is the Great Star-Cloud in Sagittarius. This lies close to the centre of Plate XXXIII, and is shewn in greater detail on Plate XXXIV. A great number of very varied investigations combine to assure us, with remarkable unanimity, that the hub of the great wheel lies somewhere in or near this star-cloud. In all probability, it lies behind the patch of dark obscuring matter which occupies the right-hand half of the plate. If so, we shall never see the hub round which the great wheel turns.

We can picture the motion of the stars most simply if we think of the path of every star as being continually curved in towards the hub of the wheel by the gravitational pull of some huge central sun. Yet it is highly improbable that any such central sun exists. If we could see through the dark clouds of obscuring matter, we should probably see nothing more than a fairly dense crowd of ordinary stars. It is most likely that the stars hold one another together by their own gravitational pulls, like the two components of a binary system, and are not controlled by a single big central mass.

As soon as we know the speeds with which the stars move round the hub, we can weigh the system of stars, just as we could weigh the sun when we knew how the planets moved round it. Each star is under the pull, not only of the stars in the hub, but of the whole system of stars, so that we can find the weight not merely of the hub but of the whole cart-wheel. And as we know that the average star has about the same weight as the sun, or perhaps somewhat less, we can say how many stars go to make up the cart-wheel.

Needless to say, we cannot tell the number with great accuracy. It is almost certainly greater than 100,000 million; that is to say, there are almost certainly more than 60 stars for each man, woman and child living on earth. It may well be twice, and perhaps even three or five times, this number.

It is not easy to realise what such numbers mean. First of all, how many stars can we see on a perfectly clear night, without a telescope at all, using only our own eyes? They look a vast multitude, and most people, if asked to guess, will say a hundred thousand, or twenty million, or some such number. But actually the very best eyesight can only see about 3000—rather more than the number of letters of type on two pages of this book.

Now imagine every one of the 3000 stars we can see to spread out into a whole new sky-full of stars. Even this feat of the imagination only gives us 9 million stars, and this is still only a tiny fraction of the whole number of stars in the sky. It is the number of letters in about forty books, each the size of this one. To imagine the total number of stars in the sky we must think of a huge library of at least half a million books, each like this. All the letters of type on all the pages of all the books of this library will be about equal in number to the stars in the sky. If we read at the rate of a page every minute for eight hours a day, it will take us 700 years to read through this library. In the same way, if we count the stars at the rate of 1500 a minute—25 a second—it will take us

700 years to count them all. Our earth is a tiny appendage to one, and a rather inconspicuous one, of this vast multitude of stars. It is less—very much less—than the dot of an "i" in our library of half a million volumes; it ought rather to be compared to a microscopic speck of dust imprisoned between two pages. And this is the speck of dust whose inhabitants thought, until about 300 years ago, that it was the centre of the universe, and that all the other stars circled round it—had in fact been created for no other purpose than to circle round it, and shed a bit of light on it now and then in the absence of the sun and moon. We begin to see now how insignificant our home in space really is, and yet, as we shall see in the next chapter, the greater part of the story remains to be told.

CHAPTER VII

OUT IN THE DEPTHS OF SPACE

We have seen how, when little was known about astronomy, it was natural to imagine that the stars went on for ever and ever, so that, however far one probed into space, one merely came to more stars. It was only like the town-bred child imagining that the lamp-posts go on for ever and ever. Yet we now know that if we go far enough out into space, we come to regions where the stars first begin to thin out, and then disappear altogether: we are now out in the depths of space beyond the Milky Way. The stars are like the lights of a vast city, but no city, however great, extends for ever, and if we go far enough, we get out of the city, and come at last to the dark open country beyond.

Yet this is not the whole story. We now know that the wheel-shaped system of stars bounded by the Milky Way is not the only system of stars in space. Far beyond the Milky Way are other cities, each with its own system of lights. The dark open country which surrounds our own city is not the end of everything; if we persevere through it for long enough, we shall come in time to another city whose lights are stars similar to those surrounding our sun. Let me explain the evidence for this statement.

When we are far out at sea we cannot see the lights of a seaside town as separate points of light; they all run together and form a sort of confused cloud of light. Then, as our ship brings us nearer inshore, we begin to see the separate lights, first the brightest and later on the fainter ones as well.

It has been the same with the distant star-cities far out in space. We have not come any nearer to them, but the ever-increasing power of our telescopes has, in a sense, brought them ever nearer to us, and, just in the last few years, we have begun to see their individual lights, and recognise them for what they are—cities of stars like our own. Their nature had, however, been suspected long before it was known with certainty. In 1755 the philosopher Kant described them as "systems of many stars, whose distance presents them in such a narrow space that the light which is individually imperceptible from each of them, reaches us, on account of their immense multitude, in a uniform pale glimmer."

As they were only seen as dim clouds of light, they were called "nebulae," a Latin word meaning mists or clouds. Not all nebulae consist of groups of stars. True nebulae are of two distinct types, which can be distinguished by their shapes. The first type are regular in shape, or very nearly so. The second type are completely irregular in shape; they form by far the more impressive objects when seen in a telescope, but only because they are nearer to us—just as the moon looks more impressive than Betelgeux. They generally look rather like drifting masses of smoke such as one sees when a house or a haystack is on fire. And, indeed, they are, so to speak, only the smoke of our own star-city, lighted up by the lights of our own star-city; they are wisps and clouds of dust and luminous gas stretching from star to star within the confines of the Milky Way, and forming light and dark patches against the sky, much as the smoke and flame of an ordinary fire form light and dark patches against the sky.

Two examples of this type of nebula, both in the constellation of Orion, have already been shewn on Plate XXVII (p. 80) and Plate XXX (p. 103). A third, in the constellation of Cygnus, is shewn on Plate XXXV.

The Great Nebulae out in Space

The other type, the nebulae of regular shape, are the distant cities of stars. They are so far away that they look singularly ineffective when viewed directly, even through a powerful telescope. Their faint light makes very little impression on our eyes. The brightest of them all, the Great Nebula in Andromeda (see Plate XXXVI), was described by the astronomer Marius as looking like "a candle-light seen through horn." To understand what these nebulae are, we must let their light impress itself, hour after hour, and perhaps even night after night, on a photographic plate. When this is done detached individual lights begin to emerge from the general light of the nebulae (see Plate XXXVII). These prove to be stars. We know that they are stars, because many of them are unmistakable Cepheid variables, shewing exactly the same characteristic and familiar fluctuations of light as the Cepheid variables nearer home. This is a great piece of good fortune, because, as we have already

PLATE XXXV

Mount Wilson Observatory

Nebulosity in Cygnus
(See p. 120.)

PLATE XXXVI

Yerkes Observatory

The Great Nebula M 31 in Andromeda

This, the most conspicuous of the star-cities in space, is slightly more distant than M 33 (Plate XXXVIII). Its light takes 900,000 years to reach us, and it is so vast that light takes nearly 50,000 years to cross from one side to the other.

PLATE XXXVII

Mount Wilson Observatory

The outer edge of the Great Nebula M 31 in Andromeda

This plate shews in detail the top left-hand corner of the nebula shewn opposite. It is seen
to consist of separate stars.

PLATE XXXVIII

Mount Wilson Observatory

The Nebula M 33 in Triangulum

Although this is the nearest of all the star-cities in space, its light takes 850,000 years to reach us. This picture would need to be magnified until it was as large as the whole of Europe before a body of the size of the sun became visible in it.

seen, we can tell the distance of any Cepheid variable from its apparent brightness or faintness. The Cepheid variables in the nebulae all appear very faint, and as we know that they are very bright stars in themselves, this alone proves that the nebulae are at very great distances.

We need rather a long yard-stick to measure this kind of distance. Light travels 11 million miles in a minute, and so about 6 million million miles in a year. Astronomers take this distance as their unit of measurement, and call it a "light-year." When the Germans speak of a distance of an hour, they mean the distance a man walks in an hour. In the same way, when an astronomer speaks of a light-year, he means the distance light travels in a year.

The nearest Star-cities

We have seen how the light from the nearest of the globular clusters takes 18,400 years to reach us, or as we can now say, the nearest globular cluster is at a distance of 18,400 light-years. But the nearest nebula, the nebula M 33[1] in the constellation Triangulum (Plate XXXVIII), proves to be about 850,000 light-years distant; it is more than forty times as remote as the nearest globular cluster.

The light by which we see the globular clusters had started on its long journey through space before man had become civilised, but that from even the nearest of the nebulae had started before man existed at all. If the first man to inhabit the earth had built

[1] M 33 means Number 33 in the catalogue of Messier.

a wireless station and sent out a call, calling all stations in space to inquire if there were any other intelligent beings in the universe, his call would not yet have reached the nearest of the nebulae.

Even the furthest of the globular clusters is less than a quarter as far from us as the nearest nebula. After we have left all the globular clusters behind, we must go four times further before we begin to encounter the nebulae. And as the globular clusters map out the confines of the Milky Way, this means that the nebulae are entirely clear of the Milky Way. If we take London to represent our own star-city in size, the nearest star-city out in space will lie somewhere near Cambridge; there is plenty of open country in between.

The second nearest star-city is only a shade further away, at a distance of 900,000 light-years. If the nearest star-city in space is represented by Cambridge, this may suitably be represented by Oxford. It is the Great Nebula in Andromeda (see Plates XXXVI, XXXVII and XLII (p. 125)), the best known of all the star-cities out in space, and the only one which is quite distinctly visible to the unaided eye. It lies almost north of the star Beta Andromedae (see Star-Map I and p. 161). As a spectacle, it must be admitted that it is very disappointing, and yet it is perhaps worth looking at once in a lifetime, if only to reflect, while you look on it, that the retina of your eye is being affected by light which has travelled uninterruptedly for 900,000 years to reach you. Waves of light, generated by the jumping about of electrons in the distant nebula 900,000 years ago,

PLATE XXXIX

Mount Wilson Observatory

The Nebula M 81 in Ursa Major

This is one of the most beautiful of the star-cities out in space, and was the first observed to be rotating. Its light takes 1,600,000 years to reach us.

PLATE XL

Mount Wilson Observatory

A Cluster of Nebulae in Coma Berenices

The majority of the objects in this photograph are nebulae, at a distance such that
their light takes 50 million years to reach us.

have been travelling undisturbed through space ever since, and now, as they enter your eye, are encountering solid substance for the first time since they left the nebula. They come in an unbroken sequence, at the rate of about 500 million million every second, and the ray of light which connects your eye up with the nebula contains enough waves to keep up the supply of waves at this rate for 900,000 years; those who like arithmetic can calculate the exact number.

Not many nebulae are near enough for Cepheid variables to be detected in them. When this can be done, the sizes and distances of the nebulae can be discovered at once, but in most cases, other methods must be used.

If a number of exactly similar objects are placed at different distances from us, they of course look of different sizes, but their surface-brightness is not affected by distance; it would be if there were dimming or obscuring matter in space, but we have every reason to think that there is so little of this that it can be disregarded except in a few special parts of the sky. Now Dr Hubble of Mount Wilson Observatory finds that nebulae of the same shape all appear to have the same surface-brightness, and differ only in apparent size. This suggests strongly that they are similar structures which differ only in being at different distances from us, so that we can tell their distances either from their apparent sizes or from the amount of light we receive from them; in brief, the smaller and fainter a nebula appears, the more distant it is. Plate XL shews a cluster of nebulae in Coma

Berenices, at a probable distance of 50 million light-years: nebulae are so thick in this part of the sky that the plate contains more nebulae than stars. Plate XLI shews a still more distant cluster of nebulae in Pegasus. Every one of the faint ill-defined objects in the plate is a nebula, there being 162 in all; many of these, could we see them from near enough, would appear as vast systems with complicated structures like those of the near nebulae shewn in Plates XXXVI, XXXVIII and XLIII (p. 126). The most distant of all the nebulae revealed by our telescopes are so remote that their light takes about 140 millions of years to reach us.

Our comparison of the Galactic System and the two nearest nebulae to London, Oxford and Cambridge proves to have been a good one in many ways. The largest telescopes reveal about two million nebulae in all, not a single one of which, so far as we can tell, is as large as our own star-city, so that, to begin with, we did well to compare this to London, the largest city in the world. Indeed many astronomers are inclined to regard the Galactic System as an aggregation of a number of overlapping star-cities, just as London is an aggregation of over-lapping towns. If London represents the Galactic System in size, Cambridge and Oxford just about represent the sizes of the two nearest star-cities. And the comparison holds good in respect of the number of inhabitants as well as of the arrangement in space; London has roughly 100 times as many inhabitants as Cambridge or Oxford, and our own star-city contains something like 100 times as many stars as

PLATE XLI

Mount Wilson Observatory

The Furthest Depths of Space

This plate shews some of the most remote objects accessible to observation—a cluster of 162 nebulae in Pegasus mostly at distances of 100 million light-years or more. Each contains enough material to make a star-city of thousands of millions of stars.

PLATE XLII

Mount Wilson Observatory

The Central Region of the Great Nebula M 31 in Andromeda

This plate shews in detail the central region of the nebula shewn in Plate XXXVI.
No stars can be discovered in the fuzzy central mass.

either of its two nearest neighbours. It may, however, seem surprising that we can speak with so
much confidence as to the total number of stars in
nebulae which are so remote that we can see only a
few of their very brightest stars.

Weighing the Star-cities

We have seen how our own system of stars, the
Galactic System, is flat like the solar system. Also,
like the solar system, it can preserve its flat shape
because it is in a state of rotation. Many of the
nebulae are also flat in shape, and it would seem
reasonable to conjecture that they, too, preserve
their flat shape through being in a state of rotation. Observation confirms this conjecture, for these
nebulae are found to be rotating. It must almost
certainly be this motion of rotation which saves the
stars on their rims from falling in towards their
centres. If we know the speed of this motion, we can
calculate the gravitational pull towards the centre,
and so can weigh the nebulae—just as, nearer home,
we weigh the sun or Jupiter or the whole Galactic
System of stars. The average nebula is found to have
two or three thousand million times the weight of the
sun.

This does not necessarily mean that there are this
number of stars in each nebula. Few, if any, of the
nebulae appear to consist solely of stars; most of
them have a central region which looks more like
a cloud of gas than a cloud of stars. At any rate
no telescope in existence can break it up into stars
(see Plate XLII). It must of course exert the same

gravitational pull as an equal weight of stars, so that the weight of this cloud of gas, or whatever it is, is included in our estimate of the weight of the nebula. But if these apparent clouds of gas do not already consist of stars, it seems likely that they are destined to form stars in due course of time. The reason why we think this is as follows.

The Evolution of the Nebulae

The two nearest nebulae, which we have compared to Cambridge and Oxford, are about as flat as pancakes. Our own system of stars, the Galactic System, is also flat, although not to quite such an extreme degree. But not all the regular-shaped nebulae are flat. Plates XLIV and XLV shew various types of observed nebulae. We see that some are as round as cricket balls. Others are a bit flattened, like an orange. Others are far more flattened, and so it goes on, until finally we come to the completely flat nebulae like the two nearest home. We can arrange the various shapes of nebulae in order of flatness, just as we could arrange a heap of beads in order of size, or colour, or shape, or by any other single characteristic we please. In Plates XLIV and XLV they have been so arranged.

Now when we have arranged the nebulae in order of flatness, we find that a number of other characteristics also change gradually as we pass along the sequence. It is as though we had threaded a heap of beads on a string, the largest at one end, the smallest at the other, and then discovered that not only the

PLATE XLIII

Mount Wilson Observatory

The Nebula M 51 in Canes Venatici

This is one of the nearest nebulae, after the two shewn in Plates XXXVI and XXXVIII, its light probably taking about 1,100,000 years to reach us.

PLATE XLIV

N.G.C. 3379

N.G.C. 221

N.G.C. 4621

N.G.C. 3115

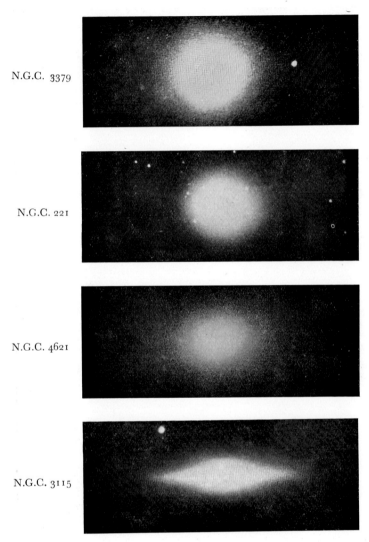

A Sequence of Nebulae—I

The sequence is continued on Plate XLV opposite.

PLATE XLV

N.G.C. 5866

N.G.C. 4594

N.G.C. 5746

N.G.C. 4565

A Sequence of Nebulae—II

Plate XLIV opposite and this plate together form a sequence of nebulae arranged in order of increasing flatness.

PLATE XLVI

N.G.C. 7217

N.G.C. 2841

N.G.C. 5457

A Sequence of Nebulae—III

Nebulae similar to the last three on Plate XLV, viewed from a different angle. The nebulae shewn in Plate XXXVIII (p. 121) and Plate XXXI (p. 104) carry on the sequence, which ends in a cloud of stars.

size, but also the shape and colour changed gradually as we passed along the string, so that in trying to arrange our beads by size we had unwittingly arranged them by colour and shape at the same time. We find, for instance, that, broadly speaking, the flattest nebulae are the largest, and *vice versa*; our arrangement by flatness is also an arrangement by size. The same is true of shape: two nebulae which have the same degree of flatness are generally almost identical in shape, and so on. In brief, nearly all of the regular-shaped nebulae can be arranged in a single sequence, like beads on a string, and all their characteristics change gradually as we pass along the string.

Or again, to go back to our old comparison, it is as though we had started to sort out a miscellaneous collection of dogs. We first arrange them by size, and then find that we have at the same time arranged them also in order of weight, height, length of coat, and so on. We should conclude that all the dogs were of one breed, and might suspect that the order of our arrangement was approximately one of increasing age.

In the same way we may say that the majority of the nebulae are of the same breed, and it seems most likely that the single sequence in which they can be arranged is simply one of increasing age, or, to be slightly more accurate, of advancing stages of development. In spite of their marked diversity of appearance, the different nebulae probably differ mainly in being more or less advanced in development, like a sequence of babies, children, boys, men and greybeards.

The Birth of the Stars

One other characteristic, which has not so far been mentioned, also changes gradually as we pass along our string of nebulae; it is, perhaps, the most important of all. One end of the string is occupied by nebulae with no flattening at all, nebulae which are as round as cricket balls. No stars can be detected in any of these: they look like mere fuzzy balls of gas or fuzzy clouds of dust. As we pass along the string the nebulae get flatter, but for a long time no stars can be detected in them. It is only after they have become very flat that stars first begin to shew. They appear first at the outer edges, the regions near the rim of the nebulae. Then as we pass along our string to still flatter nebulae, the starry regions occupy more and ever more of the nebula until finally even the centre succumbs and breaks up into stars. Indeed Dr Hubble has shewn that the sequence shewn in Plates XLIV–XLVI can be extended quite naturally by the addition, first, of our nearest neighbour in space M 33 (Plate XXXVIII, p. 121), which is nearly all stars, and then by the addition of the Lesser Magellanic Cloud (Plate XXXI, p. 104), which consists of nothing but stars. The nebula is now nothing but a cloud of stars—a star-city of the kind we have already discussed.

Thus our string or sequence of nebulae starts with something that looks like a fluffy, featureless ball of gas, and ends up as a city of stars. It would in any case be hard to resist the conjecture that such a sequence is one of advancing development, so that as we pass along it, what was originally a cloud of formless gas has been condensing into stars. We can, however, check this conjecture by calculating mathematically how a cloud of hot gas would behave as it gradually cooled down with advancing age. We find that it would go through the sequence of shapes and conditions represented on our string of nebulae, and end up as a cloud of stars. More than this, we can calculate what amount of gas ought to go to form each star—in other words, we can tell what would be the weight of stars formed in this way. Very exact calculations are out of the question, because we do not know enough about the state of the original gas, but even without this knowledge it seems fairly clear that stars formed in this conjectural way would have just about the weights which actual stars have.

This makes it highly probable that the stars are nothing but condensed drops of gas, of course on the astronomical scale, born through nebulous masses of gas condensing into detached blobs, much as a cloud of steam condenses into drops of water.

This explains very simply why stars occur in big groups—the cities of stars; each star-city is the produce of a single ball of nebulous gas. Thus we must think of the regular-shaped nebulae, not only as the dwelling-places, but also as the birth-places, of the stars. There they are born, live and die. When we arrange photographs of actual nebulae in a continuous sequence in the way I have described—globular nebulae at one end, flat ones at the other—and look along such a sequence (as we can do by studying Plates XLIV, XLV, XXXVIII, XXXI in this order), we see a chaotic mass of gas changing gradually but steadily into a multitude of stars. We are in effect studying the birth of the stars.

We discover at once why the stars all have much the same weight; it is because they are all formed by the same process. They are almost like manufactured articles turned out of the same machine.

The Evolution of the Stars

The stars do not of course permanently retain the weights they have at birth. We have already seen how they are continually losing weight, destroying their substance and turning it into radiation. Although there is plenty of divergence of opinion, most astronomers agree in thinking that the average star is born as a very fluffy baby of quite large size. Star-babies differ from human babies in that they lose size as well as weight as they grow older, and their candle-power wanes at the same time. If these views are correct, our sun is not only losing weight at the rate of 4 million tons a second, but is also shrinking in size and brilliancy. If we look far enough ahead in time we see it shrunk to an "old-man" star—possibly a "white dwarf," such as the faint companion to Sirius now is. It will not emit heat enough to prevent everything on earth freezing, so that in all probability all life will have disappeared from the earth before then.

Instead, however, of peering further into this rather dismal future, let us look backwards in time, and review the past history of our sun. First we see it merely as a baby star—a fluffier, larger and more luminous globe than it now is. Further back still, we can hardly call it a star at all: it is merely a yet more fluffy blob of gas, mixed up with other similar blobs in a nebula of fuzzy gas—the nebula which is destined ultimately to condense into our star-city. And scattered all through space are other gaseous nebulae which in the course of time will form other star-cities.

The Birth of the Nebulae

We can look even further back in time, although only rather conjecturally. Let us imagine, as a conjecture, that at the beginning of time the whole of space was uniformly filled with gas, just as a large hall or cathedral is filled with the air we breathe. Then it can be proved that the gas would not stay spread uniformly through space in this way, but would immediately begin to condense into detached balls. Once again we can calculate how much gas would go to form each ball. And the result of the calculation is significant; we find that each ball would contain just about as much gas as each of the nebulae that we believe are destined to form star-cities.

This makes it probable that the matter of the universe started as a gas uniformly diffused through space, and that the nebulae originated by the condensation of this gas. If this conjecture is sound we can piece together the story of the evolution of the universe somewhat as follows.

The History of the Universe

We start at the beginning of time, when all the atoms which were destined to form the sun and stars, the earth and the planets, your bodies and mine, and also all the radiation which has been poured out from the sun and stars ever since, are jumbled together and form a chaotic mass of gas filling the whole of space. As the gravitational pull of each bit of gas acts on all the other bits of gas, currents are gradually set up. Wherever these currents produce a slight accumulation of gas, the gravitational pull is increased, so that each small accumulation draws still more gas in towards it; Nature acts according to the law "Unto him that hath shall be given." The successful bits of gas grow into huge condensations which continually increase at the expense of the unsuccessful bits; these they ultimately swallow up into themselves. Just as the earth, sun and planets have assumed regular shapes under the action of gravitation, so these condensations now begin to assume regular shapes: they form what we have called regular-shaped nebulae. The currents of gas which started their existence now cause them to rotate, so that they are not strictly globular in shape. At first they are orange-shaped, like our rotating earth. As they shrink, their shapes continually change and they flatten out ever more and more. Again we see the gas in their outer fringes condensing into distinct masses, and the stars being born; the formless nebulae are changed into star-cities which are flat at birth and remain flat because of their rotation.

And now, as we watch the great drama enacting itself, we may notice a particular star, our sun, meeting with the unusual accident we have already described. A second star approaches it more closely than any star has ever before done, and raises tides higher than have ever been raised before—tides like huge mountains of fiery gas travelling over the sun's surface. Finally the second star comes so near that, to anyone standing on the sun, it would fill up a large part of the sky. As it does so, the gravitational pull of the second star becomes so great that the crest of the tidal wave is drawn off and itself condenses into drops. These drops are the planets, and one of the smaller of them is our earth. At first it is a chaotic mass of fiery gas, but, as it cools, its centre liquefies. In time it becomes so cool that a solid crust forms over its surface. And later, when it is cooler still, a strange new phenomenon appears on this solid crust. Groups of atoms begin to combine into coherent organisations of the kind which we— knowing nothing of their nature or of the way they first started into being—describe as life. Whatever this life is, it shews a strange capacity for reproducing itself, and as it does so, it for ever forms organisations of greater and greater complexity. Finally we see ourselves, standing at the furthest point to which time has so far unrolled itself and representing the most complex organism that has so far been produced on earth. Whether there is more complex life on other planets of other suns, or only less complex life, or perchance no life at all, we simply do not know. But as we look back down the almost endless vistas of the great corridor of time, we realise that our race is an absolutely newcomer in the universe; our brief past is a mere speck of time in the history of the universe. Still the great panorama goes on unrolling itself, and, turning our backs on the part already unrolled, we face a future stretching out to thousands and probably millions of times greater length than our past—a future longer than anything our minds can imagine. We realise that we are, in all probability, at the very beginning of the life of our race; we are still only at the dawn of a day of almost unthinkable length.

CHAPTER VIII

THE GREAT UNIVERSE

A century ago astronomy was concerned with little beyond the sun, moon and planets—the small colony we have described as the sun's family. To-day it is mainly engaged in studying in detail the various other stars and colonies of stars, such as the three stars which form the system of Alpha Centauri, our nearest neighbours in space. The aggregate of all such stars and colonies constitutes the Galactic System, the vast conglomeration of stars whose rim is the Milky Way. At the same time, astronomy has discovered that even this huge system is only one of a vast number of somewhat similar systems. The present situation may be perhaps summed up in the three statements:

(1) The earth is only one member of the sun's family.

(2) The sun's family is only one member of the Galactic System.

(3) The Galactic System is only one member of the system of star-cities in space.

This is the furthest that astronomy has travelled so far, but we may well wonder what the situation will be, say, a thousand years hence. Will the above three statements still suffice, or will they have been supplemented by more statements of the same kind? In other words, shall we find that the whole system of star-cities only forms one unit in a still vaster assembly, and this assembly perchance a mere unit in something vaster still?

The question is an old one. As far back as 1755, Kant wrote, in his *Theory of the Heavens*:

If the grandeur of a planetary world in which the earth, as a grain of sand, is scarcely perceived, fills the understanding with wonder, with what astonishment are we transported when we behold the infinite multitude of worlds and systems which fill the extension of the Milky Way! But how is this astonishment increased, when we become aware of the fact that all these immense orders of star-worlds again form but one of a number whose termination we do not know, and which perhaps, like the former, is a system inconceivably vast—and yet again but one member in a new combination of members! We see the first members of a progressive relationship of worlds and systems; and the first part of this infinite progression enables us already to recognise what must be conjectured of the whole. There is here no end but an abyss of a real immensity, in presence of which all the capability of human conception sinks exhausted.

It was a sensational conjecture, but modern science does not confirm it. Instead, it tells us that the system of star-cities constitutes the complete universe. If there is anything beyond, it can only be other complete universes, having no interaction with our own, so that the three statements on p. 135 are complete, and admit of no extension.

A Model of the Universe

We have compared the great nebulae in space to star-cities. We took London to represent our own city of stars, the system of stars in which our sun figures as a very ordinary citizen, and whose most remote members form the Milky Way. Then we saw that the two nearest star-cities in space might appropriately be compared to Cambridge and Oxford. Each inch in London or Cambridge or Oxford represents about $1\frac{1}{2}$ million million miles in the corresponding star-cities, the distance which light travels in three months. And each inch in the open country between London and Cambridge or Oxford represents the same distance in astronomical space.

In making these comparisons we were in effect building a sort of model, drawn to scale, of our own star-system and its nearest neighbours in space. The scale of the model is of course exceedingly small. It reduces the yearly path of the earth round the sun to a microscopic speck an eight-thousandth of an inch in diameter, and the whole solar system, right out to the orbit of Pluto, becomes only the size of a grain of sand. All the stars which we can see with our unaided eyes lie within a few yards of this grain of sand; most of them, indeed, within a few feet. The system of Alpha Centauri is less than 18 inches away, and Sirius less than a yard. Let us go on building our model on the same scale.

The largest telescope on earth shews about two million regular-shaped nebulae, so that about two million star-cities must be placed in our model. London, Cambridge and Oxford give quite good representations of the three we have so far discussed, but they are rather too near together for average star-cities in space. For the most part these are not quite so close as we and our nearest neighbours happen to be; we inhabit a rather thickly populated region of space. On an average, light or a wireless message takes something like two million years to travel from one star-city to the next nearest city in space. We realise what ephemeral beings we are in the universe, when we reflect that to flash a signal from one star-city to the next and get an answer back requires 60,000 times the span of a man's life.

The nebulae shewn in Plate XLIII, at a distance of 50 million light-years, must be placed in our model at about 3100 miles from London; they would be represented by a group of cities and towns somewhere in the eastern United States of America.

The furthest of the nebulae that we see in space are about 140 million light-years away—their light takes 140 million years to reach us. With our own star-city as London, and our nearest neighbour as Cambridge, these furthest star-cities of all must be placed about 8500 miles from London.

Where does this take us? A journey of 8500 miles from London over the earth's surface may take us to Cape Horn or Western Australia or mid-Polynesia or the Antarctic Continent—we can put our furthest star-cities at any of these places, and their distances from London will be about right on the scale of our model. These, together with the less distant nebulae, will now have covered almost the whole of the earth's surface. Only a small region round the Antipodes remains unoccupied—to be exact, the interior of a circle, less than 4000 miles in radius, lying in the Southern Pacific Ocean. If we are building our model on the earth's surface, we have not much room left for representing still more distant parts of space.

' Yet we must remember that the American astronomers are planning a new telescope with power to probe twice as far into space as any at present in existence; they may reasonably hope soon to be discovering nebulae twice as remote as those we have just been considering. If these new star-cities are to go into our model, we must place them 17,000 miles from London.

So long as we stay on the earth's surface, we cannot do it. It is easy enough to take a journey of 17,000 miles over the earth's surface, but this will not take us 17,000 miles from London. Indeed it will bring us nearly back to London, for we shall have gone three-quarters of the way round the world. This, you may think, shews that the earth's surface was a very bad object to choose as a model of space; we ought to have chosen something in which we can travel on for as far as we want—for ever, if the need arises.

Not only did Kant think this in 1755, as witness the passage already quoted, but most scientists would have thought the same even twenty years ago. To-day, however, we think that in one respect at least the earth's surface gives a very good model of space.

It is good just because it does not go on for ever, because it is of limited amount, because it does not contain room for endless star-cities stretching away into endless depths of space.

The Finite Universe

We have seen how, until quite recently, astronomers felt very little interest in anything beyond the sun, moon, planets, and the few nearest stars. It was not a matter of choice. Their feeble telescopes were inadequate to explore any great distance into space; whether they liked it or not, they were limited to the regions near home. They were like the Greek sailors who, three thousand years ago, explored a few small islands in the Aegean Sea. These few islands were the whole of the world for them, because they had no means of navigating to greater distances. They cared little whether the ocean round them stretched for hundreds or for thousands or for millions of miles; they could never in any case hope to get to its more remote regions.

Then men learned how to increase the size and power of their ships, as well as their own skill in the art of navigation. Longer and longer journeys were taken over the sea until, in the great times of Magellan and Drake, ships were taken round the whole world to return to their starting-point. The entire world then lay open for exploration. But, more than this, the size of the world was now known. It had been proved that the earth's surface did not extend for an infinite distance; there was only a limited amount awaiting exploration and survey, and men could hope that before very long the whole of the earth's surface would be known to them. And indeed we of to-day, four centuries later, can say that it is almost all known.

The astronomy of to-day is reaching a position similar to that of the geography of four centuries ago. The early astronomers did not concern themselves overmuch as to whether space went on for ever or not, because they knew that in any case its further stretches were as much beyond their reach as the Antipodes were beyond the reach of the early Greek sailors in the Aegean Sea. But the modern astronomer regards the universe as a finite closed space, as finite as the surface of the earth, and if he is not yet acquainted with the whole universe, he has good reason to hope that he will be before very long. We of to-day no longer think of vast unknown and unsounded depths of space, stretching interminably away from us in all directions. We are beginning to think of the universe as Columbus, and after him Magellan and Drake, thought of the earth—something enormously big, but nevertheless not infinitely big; something whose limits we can fix; something capable of being imagined and studied as a single complete whole; something capable of being circumnavigated, if you like.

And this explains why, in one respect at least, the earth's surface gives a good model of space. If we travel straight on over the earth's surface for long enough, we come back to our starting-point; we have travelled round the world. Scientists now believe that if we could travel straight on through space for long enough, we should also come back to our starting-point; we should have travelled round the universe.

The reasons for this belief are not astronomical in their nature, and it was not an astronomer who discovered that space must bend back on itself like the surface of the earth; it was Einstein, who is a mathematician and a physicist. If his theory of relativity is true, space cannot go on for ever; it must bend back on itself like the surface of the earth.

And now you will ask whether this theory of relativity is true? I cannot tell you for certain. All I can say is that every single experiment which has ever been performed to test the theory of relativity has decided in its favour. For this reason scientists to-day have no hesitation in accepting both the theory and its consequences. And one of the most important of these is that space is not of unlimited extent, but curves back on itself, and finally closes up, like the earth's surface.

Because the earth's surface bends back on itself until it closes up, there are two ways from London to New Zealand. A traveller may go east *via* Suez and the Indian Ocean, or he may go west *via* America and the Pacific Ocean. Every summer a great number of New Zealanders come from New Zealand to London, some travelling one way, some the other, so that when they meet in London, some arrive from the east, and some from the exactly opposite direction, the west. In the same way, if space is like the surface of the earth, there must be two ways from one end of the universe to the other. If we still take London to represent our own star-city, a star-city in a region of space corresponding to New Zealand will be sending out light in all directions. Some of this must fall on the earth, and we see the nebula by it. But the star-city will also be sending out light in exactly the opposite direction, and some of this, coming the other way round space, will also fall on the earth, so that we shall see the star-city by this light also. Light from the same star-city will be reaching us along two exactly opposite directions, like the New Zealanders reaching London. Consequently we shall be able to see the same star-city by looking along two diametrically opposite directions in space.

To take a definite instance, the star-city nearest us in space is the nebula M 33 in the constellation Triangulum. If light can travel round the whole of space, some of the light from this nebula ought to reach us from the direction exactly opposite to the constellation Triangulum, so that on looking in this opposite direction we ought to see the nebula M 33, although of course only as a very small and dim object, because the light by which we see it would have travelled nearly all round the universe to reach us. In the same way, if we look in the direction exactly opposite to the constellation Andromeda, we ought to see our second nearest neighbour in space, the Great Nebula in Andromeda, again as a very small and dim object.

Now when we turn our telescopes in the directions directly away from those in which our two nearest neighbours lie, we do actually see two very small dim nebulae. It has been suggested that when we look at these nebulae, we are really only looking at our two nearest neighbours the long way round space, just as a wireless listener in London might hear Daventry, exceedingly faint, by the long path round the earth, the programme having travelled over 24,000 miles round the earth before reaching his aerial. The conjecture is a fascinating one, but

I am afraid it is so improbable as to be quite un-
tenable. All the evidence goes to shew that space is
far too big for our present telescopes to see round
it, just as the earth is far too big for the ordinary
wireless receiving set to pick up programmes which
have travelled right round it.

It is important to understand that the finiteness of
space is like that of the earth's surface, not like that
of the solid earth. The solid earth is finite too, but
in a quite different way. If we travel on in a straight
line through the solid earth, we come in time to
something which is not solid earth; we have bored
a tunnel through the earth, and come out again into
open air. On the other hand, if we travel on in a
straight line over the surface of the earth, we never
come to anything which is not the surface of the
earth. Space is like that; we can never pass from
space to something which is not space.

Perhaps we shall get a better picture if we compare
space to the film of a globular soap-bubble. Then we
ourselves and all material objects which exist in
space, and all light travelling through space, must
be compared to some sort of creatures which can
only exist in a soap-film, and have no idea of stepping
sideways out of the film. Einstein's theory of relativity
shews that space is finite in the same way as the film
of a soap-bubble.

The Expanding Universe

Of late years there has been a further development
of a sensational kind. Every child knows that it is

easy to blow a soap-bubble, but far less easy to keep it in existence for more than a minute or two; after that it is apt suddenly to burst and disappear. Quite recently it has been discovered that the universe is like this. A Belgian mathematician, Lemaître, has shewn that Einstein's universe has properties like those of a soap-bubble. It is unstable, although not quite in the same way as soap-bubbles blown with a pipe. The instability of the universe takes the form of an incapacity to stand still. As soon as it comes into existence, it starts swelling out in size, and must go on expanding indefinitely. It is not so much like the soap-bubble we have blown and detached from our pipe, as like one we are still blowing; its size for ever increases, and must go on increasing to the end of time. As the soap-film increases in size, it gets ever thinner and thinner, and the different bits of it get further and further apart. So, as the universe increases in size, the distribution of the various objects in space gets more and more sparse, and the nebulae, the great star-cities which lie in the soap-film, move further and further apart. Even now, most of them are so distant that it needs a quite powerful telescope to see them; in time they will have moved even further away so that still greater telescopic power will be needed.

Indeed we have to consider an even worse case than this. For an expanding universe not only continually increases in size, but its speed of expansion for ever increases. Thus a time must come when it will be expanding so fast that no ray of light can ever get completely round it; while the light travels a million miles, the circumference of the universe will have expanded two million miles, so that the light will have a longer journey before it than it had when it started. Trying to see round the universe will be like trying to catch a train which is already moving faster than we can run. I have said such a time as this must come; I ought to add that if the calculations of the mathematicians are to be trusted, this time has already come; we entered the universe too late to see round it.

Astronomers have devices for measuring the speeds with which astronomical objects are moving away from us or towards us, so that they ought to be able to tell us whether the distant nebulae are really moving away from us, as the mathematicians assure us that they must be.

A Stampede of Nebulae

Now the results of measuring the speeds of the nebulae are absolutely sensational; they shew that practically all the nebulae are stampeding away from us at terrific rates. A runaway speed of a thousand miles a second is quite slow for a nebula; most of the nebulae seem to be receding at far greater speeds. The last nebula to be investigated at Mount Wilson Observatory was found to be receding from us at the rate of 26 million miles an hour, or about 200,000 times the speed of a fast aeroplane.

Yet, just because these apparent speeds are so enormous, many astronomers doubt whether they can be real. If they are, the whole universe must be expanding—we might almost say exploding—at a really terrific rate, judging time on the astronomical scale, and the whole universe must be a far more transitory and ephemeral phenomenon than it is usually believed to be. And the general evidence of astronomy points in exactly the opposite direction.

We can judge the ages of the stars in various ways —from their weights, their looks, their motions and so on—much as we judge the age of a horse from its teeth, its appearance and its action. So far as we can see at present, all the evidence goes to shew that the stars are millions of millions of years old. If our estimates of the ages of the stars are correct, then the universe cannot really be expanding at the terrific rate which the apparent motions of the nebulae seem to indicate. For expansion could hardly have been in progress at this rate, or anything like it, for more than a few thousand million years at most, otherwise the universe would have had to start from nothing or less than nothing.

I do not think we need mistrust the actual measurements from which the high speeds of the nebulae are deduced. Such measurements are easy to make, and are certain to be reasonably accurate. It is the principle underlying them that comes under suspicion. All sorts of things may simulate the effects produced by a high speed of recession, and it may be that one of these is responsible for the sensational appearance of speed.

Yet even if the measurements were entirely wrong, and if our whole interpretation of them were wrong —even if the supposed speeds should prove to be entirely spurious—it would still seem likely that the universe is expanding. The mathematical work of Lemaître shews that it could not stand still in any event. The only question is whether it is expanding at the terrific rate which observations on the nebulae seem at first sight to indicate, or at some slower rate. This is still an open question of a very technical kind: no doubt science will find out the truth before long. It is perhaps hardly surprising that it has not yet done so, since it was only a few years ago that it first began to contemplate the universe as a whole.

The Size of the Universe

If the universe had only just come into existence, and had not yet begun to expand to any appreciable degree, then its curvature would depend only on the distribution of matter in it. From this we can calculate that the journey of light round the universe would take about 500,000 million years.

On the other hand, if the apparent speeds of recession of the nebulae represent a real expansion of the universe, and nothing else, then the original universe, before it started to expand, must have been far smaller than this; it must have been so small that light could travel round it in about 8000 million years. The present universe, the universe after expansion, must of course be larger than this, but we can hardly say how much larger. All we know is that its circumference must be less than the 500,000 million light-years which it would be if the whole universe were not expanding at all.

Whatever may prove to be the correct interpretation of the apparent motions of the nebulae, the circumference of the universe is likely to lie somewhere between 8000 million light-years and 500,000 million light-years. The range is a wide one, and yet in a sense it matters little what the true figure is, since even the smallest permissible figure is utterly beyond the limits of our imagination. Whatever it is, the furthest distance that our telescopes have yet penetrated into space, namely, 140 million light-years, is only a very small fraction of the whole way round the universe.

The Substance of the Universe

Within this distance of 140 million light-years, about two million nebulae are visible. Each contains about as much matter as 2000 million suns, so that the total amount of matter within range of our telescopes is roughly that of 4000 million million suns. We may describe this as the total amount of matter we can see in our telescopes; the total amount of matter in the universe must of course be greater than this.

Sir Arthur Eddington has calculated that, if the nebulae are actually receding as fast as they appear to be, the total amount of matter in the whole universe must be that of 11,000 million million million suns—nearly three million times as much as we can see in our telescopes. If those parts of the universe we cannot see are essentially similar to those that we can, this can only mean that the whole universe must be nearly three million times as big as the fragment of it that we see. In this case the circumference of the

universe would be about 100,000 million light-years—
if its expansion were suddenly arrested, light could
travel round it in about 100,000 million years. But
this estimate is in any case exceedingly uncertain,
and if part of the apparent speeds of recession proves
to be spurious, so that actually the nebulae are
receding at lower speeds, then the total amount of
matter in the universe must be greater than we have
supposed, and the size of the universe also must be
correspondingly greater.

If we may judge from those parts of space which
are accessible to telescopic observation, a large part
of the matter of the universe is already condensed into
stars. We obviously cannot state the total number of
stars in the whole universe with any approach to
accuracy, but its vastness is suggested by the statement that there are probably something like as many
stars in the universe as there are grains of sand on all
the seashores of the world. Or, to take another comparison, the total number of stars in the universe is
probably about equal to the number of drops of rain
which fall on the whole of London in a day of heavy
rain. And we must remember that the average star is
something like a million times as big as the earth.

We might have supposed that space containing
such vast numbers of huge stars, would be insufferably
crowded. Quite the reverse is the case; it is emptier
than anything we can imagine. Leave only three
wasps alive in the whole of Europe and the air of
Europe will still be more crowded with wasps than
space is with stars, at any rate in those parts of the
universe with which we are acquainted.

The Age of the Universe

We cannot say anything with certainty on the age of the universe until we know the truth as to the apparent recessions of the nebulae. If these prove to be real, it will become necessary to pack all the events of astronomy, somehow or other, into a past of a few thousand millions of years. At present the whole general evidence of astronomy seems to cry out in protest against so short a past; it hardly seems possible to account for the present arrangement of the stars if their lives have been as short as this. For this reason I think it most likely that the apparent recessions of the nebulae will prove to be spurious, in which case the arrangement of the stars points to a past of millions of millions of years, and a future of similar, or even greater, length. At present the evidence looks very confused, and even contradictory, and we are far from being able to reach any final conclusion.

Whichever alternative prevails, the universe, judged on our human scales of time, is very old; the lives of men and of nations, even the whole history of mankind, are as nothing in comparison. Before man appeared on earth, the stars were much as they now are, and in all probability they will still be much the same when the last man has left the earth. The whole history of the human race is but the twinkling of an eye in comparison with the ages of the stars.

We individuals see the universe only as a traveller sees a landscape lighted up by a flash of lightning.

It was there long before the flash revealed it to us, and will be there long after darkness has closed in again. The flash is so brief that we discern no change in the landscape while it lasts, and yet we know that the landscape is not changeless. Could we illuminate it by something less transitory than a lightning flash, we should see it as an ever-changing picture of growth followed by decay. In the same way we believe that the universe is not a permanent structure. It is living its life, and travelling the road from birth to death, just as we all are. For science knows of no change except the change of growing older, and of no progress except progress to the grave. So far as our present knowledge goes, we are compelled to believe that the whole material universe is an example, on the grand scale, of this.

We have seen how the stars are continually melting away into radiation, as surely and as continuously as an iceberg melts in a warm sea. We are still in doubt as to the extent of this transformation, but there is practically no room for doubt that the sun weighs many millions of millions of tons less than it did a month ago. As the other stars are melting away in the same manner, the universe as a whole is less substantial than it was last month.

Not only is the amount of matter in the universe decreasing, but what is left continually spreads itself further and further apart. Because the sun is continually losing weight, its gravitational grip on the planets is for ever getting feebler, so that all the planets, including the earth, are continually moving further and further out from the sun into the icy cold of space. And again, all the stars of the Galactic System, right out to the Milky Way, are held together in one another's gravitational forces. As the stars turn their weight into radiation, these forces for ever weaken, with the result that the system for ever expands. Our own star-city gets ever bigger and bigger, while its individual lights get ever feebler and feebler. The same is of course true of all the other star-cities in space. Then, beyond all, we have the general expansion of the universe—the blowing-out of the soap-bubble—so that the great star-cities themselves move ever further and further away from one another. In some way the material universe appears to be passing away like a tale that is told, dissolving into nothingness like a vision. The human race, whose intelligence dates back only a single tick of the astronomical clock, could hardly hope to understand so soon what it all means. Some day perhaps we shall know: at present we can only wonder.

APPENDIX I[1]

GUIDE TO THE SKY

The two star-maps given at the end of this book will help the reader to identify the constellations and to locate stars and other astronomical objects in the sky. But the apparent motions of the stars must first be explained more fully than has yet been done. Strictly speaking it is only roughly true that the earth turns in space once every 24 hours; it is not exactly true. It is 24 hours from the instant when the sun is overhead one day to the instant when it is overhead the next day, but the earth makes a little more than one complete turn in this interval. A complete turn brings the earth back to the same position under the stars, but as the sun is itself moving forward through the constellations all the time, a little more turning is needed to bring the earth to the same position under the sun (see Fig. 2).

Fig. 2.

[1] By permission of the Syndics of the Cambridge University Press, the arrangement and much of the subject-matter of this Appendix have been taken from Sir Robert Ball's *Primer of Astronomy*. The whole has been re-written and modernised.

The sun appears to move right round the heavens once a year, so that in a complete year the sum total of all these extra bits of turning must just amount to one complete rotation. As there are $365\frac{1}{4}$ days in a year, this means that the earth makes $366\frac{1}{4}$ revolutions in $365\frac{1}{4}$ days. From this, the time of a complete revolution in space is found to be 23 hours 56 mins. 4 secs. Every day the earth spends this much time in making a complete turn in space, and then a further 3 mins. 56 secs. in catching up the motion which the sun has made through the sky in 24 hours.

Sidereal Time

If we adjust the pendulum of a 24-hour clock so that it gains 3 mins. 56 secs. every day, then the position of its hands will repeat itself every 23 hours 56 mins. 4 secs. Thus, each time the clock tells the same hour, such as 2 o'clock or any other time, the earth lies in the same direction in space, and exactly the same stars lie overhead.

Such clocks are to be found in every observatory. An ordinary clock tells us, in effect, where the *sun* is in the sky, but these clocks tell us where the *stars* are in the sky. They are known as "sidereal-time" clocks. They are started at zero-hour when the stars are in a certain agreed position, and after that shew what is known as sidereal time.

Not every one can possess a sidereal clock, but the table on p. 156 will give sidereal time accurately to the nearest hour, which generally enables us to get the orientation of the stars in space with sufficient accuracy to identify particular stars.

SIDEREAL TIME-TABLE

Local time / Date	4 p.m. to 5 p.m.	5 p.m. to 6 p.m.	6 p.m. to 7 p.m.	7 p.m. to 8 p.m.	8 p.m. to 9 p.m.	9 p.m. to 10 p.m.	10 p.m. to 11 p.m.	11 p.m. to midnight	midnight to 1 a.m.	1 a.m. to 2 a.m.	2 a.m. to 3 a.m.	3 a.m. to 4 a.m.	4 a.m. to 5 a.m.	5 a.m. to 6 a.m.	6 a.m. to 7 a.m.	7 a.m. to 8 a.m.
Jan.	0	I	II	III	IV	V	VI	VII	VIII	IX	X	XI	XII	XIII	XIV	XV
Feb.	—	III	IV	V	VI	VII	VIII	IX	X	XI	XII	XIII	XIV	XV	XVI	—
Mar.	—	V	VI	VII	VIII	IX	X	XI	XII	XIII	XIV	XV	XVI	XVII	XVIII	—
April	—	—	—	IX	X	XI	XII	XIII	XIV	XV	XVI	XVII	XVIII	—	—	—
May	—	—	—	XI	XII	XIII	XIV	XV	XVI	XVII	XVIII	XIX	XX	—	—	—
June	—	—	—	—	—	XV	XVI	XVII	XVIII	XIX	XX	—	—	—	—	—
July	—	—	—	—	—	XVII	XVIII	XIX	XX	XXI	XXII	—	—	—	—	—
Aug.	—	—	—	XVII	XVIII	XIX	XX	XXI	XXII	XXIII	0	I	II	—	—	—
Sept.	—	—	—	XIX	XX	XXI	XXII	XXIII	0	I	II	III	IV	—	—	—
Oct.	—	XIX	XX	XXI	XXII	XXIII	0	I	II	III	IV	V	VI	VII	VIII	—
Nov.	—	XXI	XXII	XXIII	0	I	II	III	IV	V	VI	VII	VIII	IX	X	—
Dec.	XXII	XXIII	0	I	II	III	IV	V	VI	VII	VIII	IX	X	XI	XII	XIII

If summer-time is in operation, the number of hours given by the table must be reduced by one.

Because the pole of the heavens always lies due north, a line drawn from it to the zenith (the point directly over our heads) and carried straight on, must finally meet the horizon at a point due south of us. This line is called the meridian. Any star we choose to select will always cross the meridian at the same place, night after night, and of course always at the same sidereal time. For instance, Sirius always crosses the meridian at sidereal time 6 hours 40 mins. at a point 106° away from the North Pole. We say that 6 hours 40 mins. is the "Right Ascension" of Sirius, and that 106° is its "North Polar Distance."

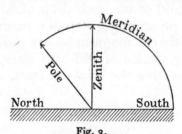

Fig. 3.

The first of the two star-maps at the end of the book shews all the bright stars which lie within 115° of the North Pole, this including all parts of the sky that can be seen easily in the latitude of England. Stars further from the North Pole than this are either for ever beyond the horizon, or too near the horizon to be seen with ease. These are shewn in the second star-map.

To discover where any given star lies at any given time, we must first find the sidereal time from the table opposite. Suppose for instance it is 7 hours.

We then know that all stars of 7 hours Right Ascension will be on the meridian. The star-map shews us which these stars are, Right Ascensions in hours being indicated round the edge of each map, and from this we can approximately locate any other star we want.

The stars shewn in these star-maps are divided into four classes, according to their apparent brightness, designated as stars of the first, second, third and fourth magnitudes. Roughly speaking, the twenty brightest stars in the sky are called stars of the first magnitude; a star which gives 40 per cent. of the light of a first-magnitude star is then called a second-magnitude star, and so on, each drop of 60 per cent. in brightness representing a step of a magnitude.

Stars of different magnitudes are indicated in the star-maps by circles of different sizes, the largest circles representing the brightest stars.

Star Regions

The star-maps divide the sky into twenty regions as follows:

NORTHERN REGIONS

Region	Name		Right Ascension		
1	Pole Star	...		All	Within 25° of pole
2	Cassiopeia	...	XXII to	II	
3	Capella	...	II ,,	VI	
4	Gemini ...	...	VI ,,	X	More than 25° and less than 70° from pole
5	Ursa Major	...	X ,,	XIV	
6	Hercules	...	XIV ,,	XVIII	
7	Vega ...	...	XVIII ,,	XXII	

EQUATORIAL REGIONS

Region	Name		Right Ascension			
8	Cetus		0	to	IV	
9	Sirius		IV	„	VIII	More than 70°
10	Regulus		VIII	„	XII	and less than
11	Arcturus ...		XII	„	XVI	110° from pole
12	Altair		XVI	„	XX	
13	Pegasus		XX	„	0	

SOUTHERN REGIONS

Region	Name		Right Ascension			
14	Fomalhaut ...		XXII	to	II	
15	Eridanus ...		II	„	VI	
16	Canopus ...		VI	„	X	More than 110°
17	Southern Cross		X	„	XIV	from pole
18	Centaur ...		XIV	„	XVIII	
19	Sagittarius ...		XVIII	„	XXII	
20	South Pole ...		All			More than 155° from pole

The principal objects of interest in these various regions are as follows:

Northern Regions

Region 1—*Pole Star* (see Plate IV, p. 15)

This region includes the whole of the constellation Ursa Minor (the Little Bear, see Plate IV), as well as parts of Cepheus, Camelopardalis, Draco (the Dragon), Cassiopeia and Ursa Major (the Great Bear). It contains little of interest except the star Polaris, or Alpha Ursae Minoris, commonly known as the Pole Star. This will easily be identified from the star-map, or from Plate IV. It may also be found by the help of the two stars known as the "pointers" in Ursa Major (see Region 5). A line drawn from Beta

Ursae Majoris to Alpha Ursae Majoris and continued about five times as far, will end near the Pole Star, which cannot be mistaken, since there is no other bright star in the vicinity.

We have already noticed (Plate IV, p. 15) that the star does not exactly coincide with the pole round which the sky appears to turn. It is about $1\frac{1}{4}°$ away, or a quarter of the distance between the "pointers," which are 5° apart. The distance is also two and a half times the diameter of the sun or moon, but this method of statement makes the distance seem unduly large since, owing to their brightness, both the sun and moon give a very deceptive impression of size. The actual pole lies on a line joining Polaris to Zeta Ursae Majoris, the last star but one in the tail of the Great Bear.

As Polaris is at a distance of some hundreds of light-years, it must be a terrifically luminous star. It is a variable star with a period of 4 days, and is accompanied by a much fainter companion.

Region 2—*Cassiopeia* (see Plate II, p. 3)

This region is mainly occupied by the constellations Cassiopeia, Andromeda, and part of Pegasus (see p. 11). The constellation of Cassiopeia lies at about the same distance from the Pole Star as the Great Bear, but in exactly the opposite direction. Its five principal stars are easily recognised, as they form a large W, the chair of Cassiopeia.

The star on the extreme right of Cassiopeia is known as Beta Cassiopeiae, or Caph; next to it comes Alpha Cassiopeiae, or Schedar. These two stars form the footpiece of the "chair."

A line drawn from Beta through Alpha and produced about four times as far brings us to the star
Gamma Andromedae, or Almak, one of the most
beautiful of all double stars. The brighter star of the
pair is yellow in colour and the smaller is bluish green;
the two stars have been compared to a topaz and an
emerald. A good telescope shews that the emerald
star is itself composed of two stars; these are found
to revolve around one another once every 55 years.
The stars are at about 400 light-years distance, and so
must be very luminous in themselves.

About half-way between Gamma Andromedae
and the nearest corner of the Great Square of
Pegasus, we find the second-magnitude star Beta
Andromedae. From this we can find the position
of a very important object, namely the Great
Nebula in Andromeda (p. 122), the only one of the
regular nebulae which can be seen clearly with the
unaided eye. It lies about a quarter of the way
from Beta Andromedae to Beta Cassiopeiae.

Region 3—*Capella* (see Plate III, p. 14)

In this region, the Milky Way sweeps through the
constellation Auriga (the Charioteer), which contains
the brilliant star Capella, or Alpha Aurigae.

Capella is easily identified, because it lies half-way
between the Belt of Orion and the Pole Star. It also
lies nearly in line with the largest side of the conspicuous quadrilateral figure which forms part of
Ursa Major, or again it may be identified by the
small V-shaped formation of three bright stars which

lies near it. These are known as Haedi, the "kids," Capella itself being the "she-goat."

Capella comes to the meridian at midnight early in December and is then only about 6° south of the zenith at London. It is the star characteristic of winter nights, just as Vega (see Region 7) is the star characteristic of summer nights. Capella is a shade less brilliant than Vega, but both are brighter than any other stars in the Northern Hemisphere. In the Southern Hemisphere, however, Sirius, Canopus, and Alpha Centauri are all more brilliant than either (see Appendix II).

Capella is a binary star, its distance, which is known with considerable accuracy, being 52 light-years. Its two components give respectively 105 and 80 times the light of the sun, and revolve around one another in 104 days. The larger star has about eleven times the diameter of the sun, and so about 1300 times its volume, yet only weighs four and one-fifth times as much as the sun. The smaller star has half the diameter of the larger, and about four-fifths of its weight. Both are yellow giants (p. 93).

Nearly on the same parallel as Capella (i.e. at the same distance from the pole) is Beta Aurigae. This again is a binary system, formed of two stars both larger than the sun, which revolve round one another in just less than 4 days, and eclipse one another in so doing, so that the light of the star shews a temporary dimming. This system is at about 100 light-years distance. Its components are of equal brightness, each giving some fifty times as much light as the sun; they are main-sequence stars of about the same physical constitution as Sirius.

South of these two stars (and about twice as far from each as they are from one another) we find another bright star, Beta Tauri, the second brightest star of the constellation Taurus (the Bull), a large part of which lies in this region. Its brightest star, Alpha Tauri, or Aldebaran, lies in Region 9, but the part which lies in Region 3 includes the famous cluster known from ancient days as the Pleiades. Even to the unaided eye these form a striking group, but they are seen to far greater advantage in a telescope of low power. They are a group of physically connected stars, all moving through space together with the same speed and in the same direction like a flock of wild birds.

If we draw a line from Beta Aurigae to Capella, and then go twice as far again, we come to Algol, or Beta Persei, the second brightest star in Perseus. This is a very famous variable star, whose variability has been known from extreme antiquity. It also is a binary system consisting of two stars, one bright and one dark, which revolve around one another every 2 days 21 hours, and eclipse one another in so doing. When the dark star comes in front of the bright one, the light suddenly begins to sink to a third of what it was, after which, without any appreciable pause, it rises again to its original value. The rise and fall occupy about 4 hours each, and the changes in brightness are easily visible to the unaided eye. To the north of Algol, across a branch of the Milky Way, lies the bright star Alpha Persei, or Mirphak.

The constellation Perseus also contains two very

fine "star-clusters" of bright stars. Both are visible to the unaided eye, looking like bright patches on the Milky Way, although of course the stars which form them are enormously nearer than the stars of the Milky Way. They are very nearly on a line joining Alpha Persei to Delta Cassiopeiae, about three-fifths of the way from the former. A small telescope shews a beautiful horse-shoe of stars in the brighter cluster, and two triangles of stars in the fainter.

Region 4—*Gemini*

Region 4 contains large parts of the constellations Gemini (the Twins), Cancer (the Crab) and the whole of Lynx (the Lynx). Its most important objects are the two brightest stars of the constellation Gemini, Alpha and Beta Geminorum, universally known as Castor and Pollux. Castor, which is probably the finest binary star in the northern sky, forms a splendid object for observation in small telescopes. One star appears about half as bright as the other; actually the two components have about 23 and 11 times the luminosity of the sun, and are at a distance of about 43 light-years. They have the general physical constitution of Sirius, with a combined weight equal to five and a half times the weight of the sun, and revolve around one another every 306 years. A third member of the colony is a faint red star, Alpha Geminorum C, which gives only a twenty-fifth part of the light of the sun, and is only visible in a good telescope.

Recently it has been discovered that every one of the three components is itself a double star, so that

Castor is really a colony of six stars. None of the three principal stars can be seen visually as a double star, even in the most powerful telescope. But spectroscopic methods similar to those used to discover the speeds of remote nebulae (p. 146) shew that each star consists of two parts moving at different speeds; thus each must be formed of two detached masses which move around one another at so small a distance apart that they cannot be seen as distinct objects in any telescope. Such stars are known as "spectroscopic binaries." The periods of revolution are 9·22 days for the brightest star, 2·93 days for the second brightest star, and only 0·814 day (i.e. 20 hours) for the faint red star. The two components of the last eclipse one another regularly as they revolve about one another; they appear to be similar stars in all respects, each having a diameter rather more than half of the sun's diameter, and a weight equal to half of the sun's weight.

Those parts of the constellation Cancer which lie in Region 4 contain no bright stars or other objects of special interest.

The constellation Lynx also contains only inconspicuous stars, but it contains many double stars and other objects interesting to the possessor of a good telescope.

Region 5—*Ursa Major*

The most conspicuous group of stars in Region 5 are the seven principal stars of the constellation of Ursa Major (the Great Bear), namely Alpha or Dubhe, Beta or Merak, Gamma or Phecka, Delta or Megrez,

Epsilon or Alioth, Zeta or Mizar and Eta or Alkaid. These form the well-known group called Charles' Wain or the Dipper. Zeta, or Mizar, is a double star which can be separated with very little telescopic assistance.

Coma Berenices (Berenice's Hair, see p. 13), also in this region, is a gathering of faint stars which is hardly dense enough to be called a cluster.

This region includes almost the whole of the constellation Canes Venatici (the Hunting Dogs), which contains the fine double star Alpha Canum Venaticorum, or Cor Caroli. The astronomer Halley gave the star its second name at the instigation of the court physician to King Charles II, who claimed that it brightened up perceptibly on the eve of the monarch's return to London. The star is easily found by drawing a line from Alpha to Gamma of Ursa Major, and then producing it one and a half times farther. A circle drawn through the three stars of the Bear's tail passes exactly through Cor Caroli. The principal star is of the third magnitude, and its companion, a third of a minute of arc away, is between the fifth and the sixth magnitude, so that it is a very easy object for a small telescope.

The constellation contains but few other stars of interest. But it contains the magnificent spiral nebula M 51 shewn in Plate XLIII (p. 126), which is commonly known as the "Whirlpool." This was discovered in Lord Rosse's giant 6 ft. reflector in 1845, and was the first nebula in which the characteristic spiral structure was observed. In small telescopes, little can be seen beyond two blobs of fuzzy light near together.

Region 6—*Hercules*

This region contains the greater parts of the constellations Hercules, Boötes (the Herdsman) and Draco (the Dragon).

Nearly midway between the stars Zeta and Eta in Hercules lies the superb globular cluster M 13 shewn in Plate XXXII (facing page 105). Although this is the most striking globular cluster in the Northern Hemisphere, it is only just visible to the unaided eye, and then only under exceptionally favourable conditions. The only globular clusters which are unmistakably visible to the unaided eye lie in the Southern Hemisphere.

Between Hercules and Boötes lie a pleasing U-shaped group of stars known as Corona Borealis (the Northern Crown). It is one of the few constellations whose name is justified by its appearance.

Region 7—*Vega* (see Plate XXXV, p. 120)

The constellation Lyra (the Lyre) in this region contains the brilliant first-magnitude star, Alpha Lyrae, or Vega. This is the brightest star in the northern sky, and is easily visible from all parts of the Northern Hemisphere, as well as from a large part of the Southern Hemisphere. As Vega is about 51° from the North Pole, it is continually above the horizon in all parts of the Northern Hemisphere whose latitude exceeds 51°, this of course including the greater part of the British Isles.

The star is very easily identified. Just as the two stars Beta and Alpha in the quadrangle of the Great Bear form pointers to the Pole Star, so the other two stars, Gamma and Delta, of the same quadrangle, form pointers to Vega. We may notice that the Pole Star, Arcturus, and Vega form an isosceles triangle.

At midnight about the end of June, Vega crosses the meridian, being about 12° south of the zenith at London. It does the same at 10 p.m. in July and at 8 p.m. in August, and so on. Thus Vega is the star of the summer night; from September until February it crosses the meridian in daylight.

In its physical constitution, Vega is similar to Sirius. It has twice the luminosity of Sirius, or about fifty times that of the sun. It is at a distance of about 26 light-years.

The constellation Lyra also contains the double star Epsilon Lyrae. As the two components are a twentieth of a degree apart, good eyesight alone suffices to see them separately on a dark, clear night, although an opera-glass or a small binocular helps. A quite small telescope shews that each of the two constituents is itself a double star.

This region includes the whole of the constellation Cygnus. This contains the first-magnitude star Deneb, or Alpha Cygni, and also Albireo, or Beta Cygni, an exquisite double star whose components are of contrasted colours, lying at the beak of the long outstretched neck of the Swan. This region contains some of the richest parts of the Milky Way.

Equatorial Regions

We now come to the six equatorial regions, which figure in both Star-Maps I and II.

Region 8—*Cetus*

From the sidereal time-table, we find that Region 8 is on the meridian about 6 p.m. in January, so that it can be well seen in the winter months soon after darkness sets in. Towards the end of August it crosses the meridian at 4 a.m.; in September at 2 a.m.; in October at midnight; in November at 10 p.m. and in December at 8 p.m. Primarily, then, it is a region for observation on autumn evenings.

Although Cetus is said to be the largest constellation of all, it does not possess many bright stars. It contains two second-magnitude stars and nine of the third and fourth magnitudes.

The variable star Omicron Ceti (p. 92), or Mira Ceti, lies in this constellation. The German astronomer Fabricius discovered its variability more than 300 years ago. Its light continually changes, with a period of about eleven months, and exhibits extraordinary fluctuations of brightness. From being a faint telescopic star of the ninth magnitude, Mira slowly brightens up to the eighth, seventh, sixth magnitude in succession, after which it becomes visible to the unaided eye and gradually ascends to the second magnitude. It reaches its brightest about four months after the rise commences, and after staying at its brightest for nearly a month, it begins to decline. Slowly its brightness wanes until, five months later, it is again the insignificant telescopic star of the ninth magnitude from which it started. It is justly called *Mira* Ceti, the wonderstar of Cetus, for the light that it yields when at its maximum is about 500 times that given when at its minimum.

Observers will do well to notice a gigantic **W**, of which Alpha Ceti (Men Kar) and Alpha Arietis (Hamal) form the lower points, while Alpha Tauri (Aldebaran), the Pleiades and Beta Persei (Algol) mark the upper points.

Region 9—*Sirius* (see Plate III, p. 14)

This is an exceptionally interesting part of the sky, containing the group of constellations which represent Orion surrounded by animals (see p. 12). It includes the whole of Orion and Canis Minor (the Little Dog) as well as large parts of Canis Major (the Big Dog), Taurus (the Bull), Lepus (the Hare), and Monoceros (the Unicorn).

The brightest star of the constellation Canis Major is Alpha Canis Majoris, or Sirius, the brightest star in the sky. This lies in the Southern Hemisphere, but, as it is only 18° south of the equator, it can be seen at suitable times from all parts of the earth except for a small region inside the Arctic Circle. It crosses the meridian at midnight about New Year, so that in our northern latitude it is best seen in spring evenings or autumn nights after midnight. It would be a fascinating star even if it were only for its beautiful scintillation of colours. Actually it is a typical white star, but as a result of its twinkling, it appears to flash forth varied tints in rapid succession.

From the time of Homer down to the present day, Sirius has been called the Dog Star, and it is indicated by a dog in various Egyptian monuments. Its rising at the same time as the sun in midsummer was supposed to indicate the beginning of the inundations of the Nile.

Except for this one bright star, the constellation of Canis Major contains but little of interest. The most characteristic feature of Region 9 is the constellation of Orion, of which a photograph is given in Plate XXVIII (p. 81). About 10 o'clock on a January night, an observer in the Northern Hemisphere will have the constellation of Orion directly to the south of him. A little to the right of the Belt of Orion and below it is the first-magnitude star Rigel, or Beta Orionis, a star whose candle-power is known, with considerable accuracy, to be about 15,000 times that of the sun. An equal distance to the left of the Belt and above it is the equally interesting star Betelgeux, a red giant with about 300 times the diameter of the sun, and 1200 times its candle-power. These two brilliant stars, together with Sirius and Alpha Tauri, or Aldebaran, form a very conspicuous lozenge of first-magnitude stars (see Appendix II). The Belt of Orion lies almost exactly at the centre of this lozenge. A line drawn through the three stars of the Belt and produced to eight times the length of the Belt in each direction will have Aldebaran at its northern, and Sirius at its southern, end.

Exactly below the middle star of the Belt of Orion

is the Swordhandle (Plate III, p. 14), which contains the great nebula in Orion, one of the most interesting of all objects for the telescope (see Plate XXVII, p. 80).

This region also contains the constellation Canis Minor, whose brightest star is Procyon. Its position is easily found by carrying on to the left the line of the two stars Gamma Orionis (Bellatrix) and Alpha Orionis (Betelgeux) which form the uppermost side of the quadrangle of Orion. Castor and Pollux, the two brightest stars of the constellation Gemini, lie almost exactly on a line joining Procyon to the Pole Star.

Region 10—*Regulus*

This region contains large parts of the constellations Leo (the Lion), Crater (the Cup), Hydra (the Water-snake), and Cancer (the Crab). The pointers in the Great Bear which we use to find the Pole Star will also find the constellation Leo for us. For the line through the pointers which leads to the Pole Star in one direction leads to Leo in the other, Leo being about as far on one side of the pointers as the Pole Star is on the other side. The stars of this constellation form a remarkable configuration. The brightest of the group, the first-magnitude star Alpha Leonis, or Regulus, is the first of a notable curve of stars which forms the head of the Lion, and is sometimes called the "Sickle." The rest of the constellation lies on the convex side of the Sickle, and terminates in the second-magnitude star Beta Leonis, or Denebola, at the end of the Lion's tail.

The second brightest star in the Sickle, Gamma

Leonis, is a double star, well seen in a small telescope. The brighter component is of the second magnitude and the fainter, only 3 seconds of arc away, is of the fourth. The two stars shew an interesting contrast of colour. It is useful to notice that Aldebaran, Gamma Geminorum, Gamma Leonis, and Denebola are almost in line with one another.

In the centre of the Sickle is the direction from which a magnificent shower of shooting-stars arrived on the night of November 13–14, 1866. This shower returns with more or less regularity, and whenever it does so it is referred to as a display of the Leonids (see p. 65).

The constellation Cancer contains no brilliant stars, but is characterised by a curious star-cluster known as Praesepe, or the Beehive. To the unaided eye it appears merely as a hazy looking spot between Gemini and Leo. With the slightest optical assistance, such as is afforded by an opera-glass or a small pair of binoculars, Praesepe is resolved into stars.

Region 11—*Arcturus*

This region contains large parts of the constellations Virgo (the Virgin), Serpens (the Serpent), and Libra (the Balance), as well as a small part of the constellation Boötes (Region 6) which includes its brightest star Arcturus, or Alpha Boötis.

This is the brightest star in the Northern Hemisphere after Vega and Capella. It is very easily identified. We find Ursa Major, the Great Bear, we follow the sweep of its tail for a little less than twice its length, and we come to Arcturus.

Arcturus comes on the meridian at midnight in
the latter part of April, and is then about 30° south
of the zenith to an observer in England. Being only
19° north of the equator, it is visible from everywhere on the earth's surface except the interior of
the Antarctic Circle.

The constellation Virgo has as its most conspicuous
feature the first-magnitude star Alpha Virginis, or
Spica. If a line drawn from Alpha Ursae Majoris to
Gamma Ursae Majoris be prolonged with a slight
curve it will lead to Spica. It will be noticed that
three fine stars, Arcturus, Spica, and Denebola, form
a triangle whose three sides are very nearly equal.

The constellation Serpens, which also lies in this
region, can be identified by its brightest star Alpha
Serpentis, which lies to the left of Arcturus.

The constellation Corvus lies a little below Spica to
the right, its two bright stars, Beta Corvi and Gamma
Corvi, forming a V-shaped triangle with Spica.

Region 12—*Altair* (see Plate XXXIII, p. 114)

This interesting part of the sky contains large parts
of the constellations Aquila (the Eagle), Serpens
(the Serpent), Ophiuchus (the Serpent-holder),
Sagittarius (the Archer), and Sagitta (the Arrow).
It also includes that part of the constellation Hercules
which contains the star Alpha Herculis, a beautiful
double star with components of finely contrasted
colours, orange and bluish green.

The constellation Aquila is distinguished by the
first-magnitude star Altair, or Alpha Aquilae. We
may notice the striking triangle formed by the three

stars Altair, Vega and Alpha Cygni, or Deneb (which should not be confounded with Denebola in the tail of Leo). A line drawn from Vega under Beta Cygni passes near a line of three stars of which Altair, the middle star, is by far the most magnificent. This line of stars forms a conspicuous feature of Aquila, so much so that it is occasionally mistaken for the Belt of Orion.

The Milky Way passes across part of the constellation Aquila, legend asserting that at this point the celestial eagle is flying across the celestial river termed the Milky Way.

The three leading stars of Ophiuchus, together with Alpha Herculis, form an irregular quadrangle whose centre is at about the same distance from the pole as Altair.

Region 13—*Pegasus*

This region contains the constellations Aquarius (the Water-carrier), Pisces (the Fishes), Capricornus (the Goat) and other smaller constellations. Also, in conjunction with Regions 2 and 8, it contains the Great Square of Pegasus, which almost rivals the Great Bear and the Belt of Orion as a familiar feature of the sky. It is formed of the three brightest stars of the constellation Pegasus, Alpha, Beta, and Gamma Pegasi, together with a fourth star, Alpha Andromedae, which belongs to the neighbouring constellation of Andromeda.

A good test of eyesight and "good seeing" is to count how many stars inside the Square of Pegasus can be seen by the unaided eye. It is seldom that more than 30 can be counted in Great Britain, but the number increases as we pass to the clearer skies of the south. As many as 102 have been counted in Athens.

Southern Regions

We now pass to regions which are so far south that their greater parts can never be observed from Great Britain.

Region 14—*Fomalhaut*

This region is one of the brightest parts of the southern sky. It contains two first-magnitude stars, Alpha Eridani, or Achernar, and Alpha Piscis Australis, or Fomalhaut.

Piscis Australis, lying to the south of Pisces and Aquarius, is a small collection of stars containing Fomalhaut as its one conspicuous object. The line from Fomalhaut to Achernar carried on for an equal distance in the same direction brings us to the brilliant star Canopus, the brightest in the whole sky after Sirius. We have here three first-magnitude stars lying exactly in line. This line is of great help to observers in the Southern Hemisphere in identifying groups of southern stars. Only the most northerly of the three, namely Fomalhaut, is visible from Great Britain.

Region 15—*Eridanus*

The chief interest in Region 15 is provided by the long celestial river Eridanus. According to the ancient division of the constellations, this had its origin at

Achernar ("The End of the River") from which it flowed away to the north through a succession of bright stars. It first passes an interesting group of four stars of the fourth and fifth magnitudes. A little further on it comes to a star of the third magnitude, and then, winding still further north, it finally enters the Equatorial Region 8.

Subsequently the constellation was extended further to the south, so that the river now flows also to the south of Achernar, into the constellation Hydrus (Region 20).

Eridanus, which is one of the largest constellations in the sky, contains nearly 300 stars which are visible to the unaided eye. Yet none of these except Achernar is brighter than the third magnitude.

Region 16—*Canopus*

The famous constellation Argo Navis, or, more simply, Argo, is the characteristic feature of Region 16. It is so large that it is usually found convenient to subdivide it into three smaller constellations— Carina (the Keel), Puppis (the Stern), and Vela (the Sails).

The brightest star of the whole constellation, Alpha Argus, or Canopus, is second only to Sirius in brilliance. But whereas Sirius is fairly near to us, Canopus, which looks only a little less brilliant, is known to be enormously remote, and so must be incomparably brighter in itself. Unfortunately neither its distance nor its candle-power are known with any accuracy.

Region 17—*Southern Cross* (see Plate XXIX, p. 102)

This region contains two of the most remarkable constellations in the southern sky, namely Centaurus and Crux, or the Southern Cross.

The Southern Cross, containing a number of bright stars in a comparatively small region, is popularly supposed to be as characteristic of the southern sky as the Great Bear is of the northern.

The line which forms the long dimension of the Southern Cross points nearly to the South Pole in one direction, and across Centaurus to Beta Corvi in the other. The shorter dimension of the Cross points to the two brightest stars of the Centaur, to which we shall return in Region 18.

The brightest star of the Cross, Alpha Crucis, is that nearest to the South Pole. The second brightest is the most easterly star, Beta Crucis. Close to it is an eighth-magnitude star which Sir John Herschel described as "the fullest and deepest maroon red, the most intense blood-red of any star I have seen. It is like a drop of blood when contrasted with the whiteness of Beta Crucis."

This region contains one of the most brilliant parts of the Milky Way (see Plate XXIX, p. 102), and also one of its most remarkable features, a pear-shaped black patch on the sky, 8° long by 5° wide, which the early navigators and astronomers called the Coal Sack. Early Australian folk-lore interprets this as a yawning pit of darkness, and also as the embodiment of evil in the shape of an emu, which lies in wait at the foot of a tree represented by the Stars of the Cross for an opossum driven by its persecutors to take refuge among its branches. We know now that the Coal Sack is not a hole at all, but a cloud of dark matter which blots out the stars behind it (see p. 102).

Centaurus is not only of vast extent, but also contains a greater number of brilliant stars than any other constellation. It has two stars of the first magnitude, one of the second, five of the third, seven of the fourth and no fewer than thirty-nine of the fifth.

Region 18—*Centaur* (see Plate XXIX, p. 102)

Alpha Centauri, the brightest star in the constellation of the Centaur, is within 30° of the South Pole, so that in the Northern Hemisphere it can hardly be seen except from the tropics.

It is easily identified from the fact that another almost equally splendid star, Beta Centauri, lies within 5° of it. Such a juxtaposition of two first-magnitude stars is not found anywhere else in the sky. Castor and Pollux are separated by about the same distance, namely 5°, but are not equal in brightness to Alpha and Beta Centauri.

This region also contains the fine constellation Scorpio, whose brightest star, Antares, or Alpha Scorpii, lies near the end of a chain of second- and third-magnitude stars. This constellation can just be seen in the summer months from the latitudes of Great Britain, and contains some of the richest fields of the Milky Way. Of all the conspicuous stars in the sky, Antares appears the reddest, being followed by Betelgeux and then by Aldebaran. These three stars are all red giants, Antares having about 450 times the diameter of the sun, Betelgeux 300 times and Aldebaran 40 times.

Region 19—*Sagittarius* (see Plates XXXIII, XXXIV, pp. 114, 115)

The most conspicuous stars in Region 19 are two second-magnitude stars. The first is Alpha Pavonis, the brightest star of the constellation Pavo; the second is Alpha Gruis, the brightest star of the constellation Grus, which lies mainly in Region 14.

The Milky Way is specially rich and beautiful in this region.

Region 20—*South Pole*

There is no star to identify the position of the South Pole in the way in which the Pole Star fixes that of the North Pole. The most remarkable objects in Region 20 are the Greater and Lesser Magellanic Clouds, Nubecula Major and Nubecula Minor (see Plate XXXI, p. 104). Even to the unaided eye they are conspicuous objects, the Greater Cloud remaining visible even in full moonlight.

On the border of the Lesser Magellanic Cloud lies one of the nearest of the star-clusters, 47 Tucanae; this again is easily visible to the unaided eye.

THE TWENTY APPARENTLY BRIGHTEST STARS

Star	Constellation	Distance in light-years	Luminosity (compared to sun)	Star-maps (region)	See pages
Sirius (Sparkling)	α Canis Majoris	8·6	26·3	9	77, 97, 170
Canopus	α Carinae	*	*	16	177
Alpha Centauri	α Centauri	4·3	1·3	18	97, 179
Vega	α Lyrae	26	50	7	162, 167, 168
Capella (The she-goat)	α Aurigae	52	185	3	161
Arcturus	α Boötis	41	100	11	173, 174
Rigel	β Orionis	500	15,000	9	171
Procyon (Preceding the Dog)	α Canis Minoris	10·5	5·5	9	77, 97, 172
Achernar (End of the River)	α Eridani	70	200	14	177
Beta Centauri	β Centauri	300	3,000	18	179
Altair	α Aquilae	16	9·2	12	174
Betelgeux (The arm-pit)	α Orionis	200	1,200	9	81, 92, 171
Alpha Crucis	α Crucis	230	1,600	17	178
Aldebaran (The hindmost)	α Tauri	57	90	9	163, 171
Pollux	α Geminorum	32	28	4	164
Spica (Ear of Wheat)	α Virginis	230	1,500	11	174
Antares (Rival of Mars)	α Scorpii	380	4,000	18	179, 180
Fomalhaut (Mouth of the Fish)	α Piscis Australis	24	13·5	14	176
Deneb	α Cygni	600 (?)	10,000 (?)	12	175
Regulus (The Ruler)	α Leonis	56	70	10	172

* Unknown, see p. 177.

THE PLANETS

Planet	No. of satellites	Size compared to the earth			Distance from sun (compared to earth)	Period of revolution (years)	Speed in orbit (miles per sec.)	See pages
		Diameter	Volume	Weight				
Mercury	0	0·39	0·06	0·04	0·39	0·24	29·7	34, 51
Venus	0	0·97	0·92	0·81	0·72	0·62	21·7	34, 52, 54
Earth	1	1·00	1·00	1·00	1·00	1·00	18·5	53, 54, 64
Mars	2	0·53	0·15	0·11	1·52	1·88	15·0	55, 58
Asteroids	—	—	—	—	1·46–5·71	1·76–13·7	—	64
Jupiter	9	10·95	1312	317	5·20	11·86	8·1	56, 57, 73
Saturn	9	9·02	734	95	9·54	29·46	6·0	62
Uranus	4	4·00	64	14·7	19·19	84·01	4·2	19, 74
Neptune	1	3·92	60	17·2	30·07	164·78	3·4	19, 74
Pluto	—	*	*	*	39·8	248	2·9	21, 75

* The diameter, volume and weight of Pluto are still uncertain, but are probably all somewhat less than those of the earth.

THE MOTION OF THE PLANETS

(The table gives the approximate times at which the planets are furthest from the sun. At these times Mercury and Venus are most easily seen, while Mars, Jupiter and Saturn, being exactly opposite the sun, cross the meridian at midnight and so appear at their best in the night sky.)

Year	Mercury As morning star	Mercury As evening star	Venus As morning star	Venus As evening star	Mars	Jupiter	Saturn
1931	Feb., May, Sept.	Apr., Aug., Nov.	Feb.	—	Jan.	Jan.	July
1932	Jan., May, Sept., Dec.	Mar., July, Nov.	Sept.	Apr., Nov.	—	Feb.	July
1933	Apr., Aug., Dec.	Mar., July, Oct.	—	Nov.	Mar.	Mar.	Aug.
1934	Apr., July, Nov.	Feb., June, Oct.	Apr., Nov.	—	—	Apr.	Aug.
1935	Mar., July, Oct.	Feb., May, Sept.	—	July	Apr.	May	Sept.
1936	Feb., June, Oct.	Jan., May, Sept., Dec.	June	—	—	June	Sept.
1937	Feb., June, Sept.	Apr., Aug., Dec.	—	Feb.	May	July	Oct.
1938	Jan., May, Sept.	Apr., July, Nov.	Feb.	Sept.	—	Aug.	Oct.
1939	Jan., May, Aug., Dec.	Mar., July, Oct.	Sept.	—	July	Sept.	Nov.
1940	Apr., Aug., Dec.	Feb., June, Oct.	—	Apr.	—	Nov.	Nov.
1941	Apr., July, Nov.	Feb., June, Sept.	Apr., Nov.	Nov.	Sept.	Dec.	Nov.
1942	Mar., July, Oct.	Jan., May, Sept.	—	—	—	—	Dec.
1943	Feb., June, Oct.	Jan., Apr., Aug., Dec.	June	June	Dec.	Jan.	Dec.
1944	Feb., June, Sept.	Apr., Aug., Dec.	—	—	—	Feb.	Dec.
1945	Jan., May, Sept.	Mar., July, Nov.	June	Feb.	—	Mar.	—

INDEX

CAMBRIDGE: PRINTED BY W. LEWIS, M.A., AT THE UNIVERSITY PRESS